Adaptive Governance and Water Conflict

New Institutions for Collaborative Planning

EDITED BY

John T. Scholz
Bruce Stiftel

RESOURCES FOR THE FUTURE
Washington, DC, USA

Printed in the United States of America

An RFF Press book
Published by Resources for the Future
1616 P Street NW
Washington, DC 20036–1400
USA
www.rffpress.org

Library of Congress Cataloging-in-Publication Data

Adaptive Governance and Water Conflict: New Institutions for Collaborative Planning/ John T. Scholz and Bruce Stiftel, editors.
p. cm.
Includes bibliographical references and index.
ISBN 1-933115-18-1 (hardcover : alk. paper) — ISBN 1-933115-19-X (pbk. : alk. paper)
1. Natural resources—Management. 2. Sustainable development. 3. Environmental policy. I. Scholz, John T. II. Stiftel, Bruce.
HC85.S33 2005
333.7--dc22 2004030206

The paper in this book meets the guidelines for permanence and durability of the Committee on Production Guidelines for Book Longevity of the Council on Library Resources. This book was typeset by Peter Lindeman. It was copyedited by Steven Jent. The cover was designed by Rosenbohm Graphic Design. Cover photo provided by South Florida Water Management District.

ISBN 1-933115-18-1 (cloth) ISBN 1-933115-19-X (paper)

About Resources for the Future *and* RFF Press

RESOURCES FOR THE FUTURE (RFF) improves environmental and natural resource policymaking worldwide through independent social science research of the highest caliber. Founded in 1952, RFF pioneered the application of economics as a tool for developing more effective policy about the use and conservation of natural resources. Its scholars continue to employ social science methods to analyze critical issues concerning pollution control, energy policy, land and water use, hazardous waste, climate change, biodiversity, and the environmental challenges of developing countries.

RFF PRESS supports the mission of RFF by publishing book-length works that present a broad range of approaches to the study of natural resources and the environment. Its authors and editors include RFF staff, researchers from the larger academic and policy communities, and journalists. Audiences for publications by RFF Press include all of the participants in the policymaking process—scholars, the media, advocacy groups, NGOs, professionals in business and government, and the public.

Dedication

To those resolving the daily challenges of adaptive governance

Contents

Preface

WATER POLICY SEEMS in perpetual crisis. The water conflicts familiar to the arid Western states are increasingly commonplace even in the water-rich East. For example, an unresolved "water war" pits Florida against Georgia and Alabama over waters of the Apalachicola/Chattahoochee/Flint system, while the cities and counties surrounding Tampa Bay have fought for years to decide how to secure sufficient supplies for their near-term future. New conflicts extend beyond the statutory authority, competence, geographical jurisdictions, and political constituencies of highly specialized water authorities. Everglades' restoration has strained Florida's relationship with the federal government for a decade, and involves dozens of federal, state, and local authorities ranging from water management districts, agricultural and transportation agencies to the Environmental Protection Agency, the U.S. Army Corps of Engineers, the National Park Service, and many others. Increasingly, water demands for population and economic growth confront limitations imposed by poorly understood natural systems, even in states where rainfall exceeds four feet each year.

Some see the answer in technology, and indeed technological fixes have come to the rescue in quite a few instances. Others think we can craft solutions through pork barrel trades, and sometimes we do. To us, however, the incessant cry of crisis suggests structural deficiencies. Our fragmented federal system of partisan mutual adjustment governing a Keynesian-style regulated market economy seems to give insufficient priority to a life-giving renewable natural resource whose price is kept artificially low. Overconsumption is encouraged; efficiency is not particularly prized; externalities abound. Institutions responding to these problems fight an uphill battle to meld diverse statutory authorities and to cross jurisdictional boundaries amidst scientific uncertainty and rapid technological change.

The Water Governance Project at Florida State University began with the belief that the new water conflicts provide important lessons about institutional changes necessary to adapt existing water policies and water management institutions to the needs of the twenty-first century. By studying the successes and failures of the innovative processes that evolved to deal with conflicts, we sought a clearer understanding of the challenges of *Adaptive Governance*—the kind of governance structure that can both preserve the strengths of existing specialized authorities to exploit natural resources and explore alternatives in order to ensure the sustainability of both human and natural systems. Our plan was to research the recent history of a sample of Florida's intractable water conflicts and analyze these experiences through a wide variety of theoretical and practical lenses.

The two editors brought different intellectual backgrounds to this work: Scholz, a political scientist; Stiftel, an urban planner. We quickly assembled a larger team that included administrators, engineers, lawyers and policy scientists, as well as other political scientists and planners. The interdisciplinary mix has caused strife at times, but in the end, we think the product is much richer and more than a little wiser for the debates we've had.

In the spring of 2003, eight investigators prepared studies of water conflicts in Florida. What quickly emerged when we compared these cases to previous studies of environmental conflicts was that Florida's conflicts had much in common with water conflicts throughout the nation, and that the innovative, adaptive governance structures that emerged had much to say to the nation about the governance of other natural resources and environmental issues as well. These cases provided the background for analysis of Florida's water policy system by an insightful group of policy leaders and nationally prominent scholars who met in a public conference in Tallahassee in November. Revisions of the case studies and papers presented at that conference led to the chapters in this book.

Our intent is to include in one volume a set of cases and analyses that will be valuable to water professionals who are searching for new directions to reform policymaking structures that no longer work; to policy scientists who work to understand the dynamics of natural resource decisionmaking and management; and to advanced undergraduate and graduate students with environmental interests in economics, engineering, geography, law, political science, public administration, and urban planning. By blending case histories with analytical essays, scholarly with policy-oriented analyses, we hope to reach student, scholar, and professional alike in ways not often possible in works that present either experience, theory, or pragmatic analysis.

We seek to fill a gap in the literature of environmental policies and the governance of natural resources—how to govern conflicts that extend beyond the authority, competence, and interest of existing authorities, but cannot be readily resolved by political or judicial institutions. The policy literature provides useful background for environmental policies, politics, and the relevant institutions of governance; adaptive management and conflict resolution provide insight into conflicts and decision processes to resolve them. Our task is to integrate both perspectives to understand the role of emergent institutions of adaptive government in the American federalist system.

For teaching purposes, our case studies focus on Florida water conflicts to provide a context that is limited enough for students to master yet multifaceted enough to encompass the issues important for understanding adaptive governance. Each case tells the general story of one selected conflict, beginning with the origins and institutional settings and focusing on the innovative institutional arrangements—some successful, some not—that evolved to grapple with five challenges to adaptive governance as outlined in the introduction.

The subsequent analyses by practitioners and researchers view the case studies from different experience-based and theoretical perspectives, all written for an educated general audience. Each analyst highlights different challenges and different solutions; some emphasize the limitations imposed by the existing institutional structures, some expand on collaborative decisionmaking as applied to these conflicts, and some clarify a particular perspective for analyzing these processes. The final chapter provides a brief summary and assesses the current conditions for the theory and practice of adaptive governance. Since the cases and analyses cover multiple topics, they could be assigned for course readings in many ways. The entire book could be used for a 4-5 week segment of an undergraduate course on government agencies and public policy; beginning with the introduction and institutional framework (Chapter 1), then using cases and analyses for each of the five challenges: 7, 8, 12, 14, 16 for *Representation*, 3, 5, 6, 14, 20, 21 for *Process Design*, 6, 9, 18, 19 for *Scientific Learning*, 2, 7, 13, 15, 17 for *Public Learning*, and 4, 9, 10, and 11 for *Problem Responsiveness*.

This work would not be the same without the insights of Ayşın Dedekorkut, who piloted the first of the case studies and was instrumental in three others. The initial selection of cases and the organization of the conference would not have been possible without the cooperation of FSU colleagues Andrew Dzurik (Civil and Environmental Engineering), Richard Feiock (Public Administration), David Markell and J.B. Ruhl (Law), and David Rasmussen, then director of FSU's DeVoe Moore Center for Critical Issues in Economic Policy and Government, who saw value in supporting this work with endowment funds from the Center. We are indebted to Florida State's Jeff Dickey, Keith Ihlanfeldt, Susan Ihlanfeldt, Mike McDaniel, Jacqueline Martin, Sherry Rice, and Toni Sturdevant; Grace Hill, Steven Jent, and Don Reisman of Resources for the Future Press; Bruce Ritchie of the Tallahassee *Democrat*; Stuart Langton and David Moore of the Southwest Florida Water Management District; and Terrance Salt of the U.S. Department of the Interior. The errors, of course, are our own.

John T. Scholz
Bruce Stiftel

Contributors

SIMON A. ANDREW studies public administration and policy at Florida State University. He graduated with an MA in Development Economics from the School of Development Studies at University of East Anglia (UK) and a Masters in Public Administration from Texas A&M University, Corpus Christi.

RAMIRO BERARDO studies political science at Florida State University with support from the Fulbright program. He holds a degree in political science from the Catholic University of Córdoba (Argentina). He was chief editor of *Ciudadanos*, a monthly publication for the study of social movements.

MICHAEL R. BOSWELL is associate professor of City and Regional Planning at California Polytechnic State University, San Luis Obispo, where he teaches courses in policy analysis, environmental planning, and planning theory. His professional experience includes work as an environmental planner for Brevard County, Florida and the Florida Department of Environmental Protection.

AYŞIN DEDEKORKUT is assistant professor of City and Regional Planning at Izmir Institute of Technology (Turkey), where she specializes in natural resource management. Her current research focuses in the area of water resources management and collaborative planning, consensus building, and alternative dispute resolution.

JOHN FORESTER is professor of City and Regional Planning at Cornell University. He studies the micro-politics and ethics of planning practice, including the ways planners work in the face of power and conflict. His most recent book is *Israeli Planners and Designers: Profiles of Community Builders* (co-edited with Raphaël Fischler and Deborah Shmueli), a collection of oral histories.

RICHARD HAMANN is an associate in law with the University of Florida, where he specializes in water law, environmental law, and environmentally sensitive ecosystems. He has conducted research and published on a wide variety of environmental, land use and water management issues, including the management of large-scale ecosystems for ecological sustainability.

ROBERT M. JONES is director of the Florida Conflict Resolution Consortium, where he works with representatives from state and local government to design and implement collaborative approaches to planning and dispute resolution. A member of the California Bar and a graduate of the University of California, Davis, School of Law, he has served on the Commission on Qualifications of the Society for Professionals in Dispute Resolution, on the Florida Supreme Court's committee on mediator training, and currently on the American Bar Association, Ad Hoc Committee on Federal Government ADR Confidentiality.

STEVEN LEITMAN is a planning consultant with the Damayan Water Project. A graduate of Florida State University's urban and regional planning program, he has thirty years of experience with conservation and water management in the Apalachicola River watershed.

MARK LUBELL is assistant professor in the Department of Environmental Science and Policy at the University of California, Davis. His research focuses on collective action problems, including watershed management and environmental activism. He holds a doctorate in political science from the State University of New York at Stony Brook.

CONNIE P. OZAWA is professor of Urban Studies and Planning at Portland State University, and conducts research on collaborative processes in environmental policy and transportation and land use planning. She is author of *Recasting Science: Consensus-Based Procedures in Public Policy Making*. She holds a Ph.D. in urban planning from the Massachusetts Institute of Technology.

DONALD J. POLMANN is director of Science & Engineering at Tampa Bay Water, where he manages a team in the support of regional public water supply through water resource planning and permitting, engineering studies and evaluations, environmental monitoring and assessment, laboratory services, and optimized resource management.

PAUL J. QUIRK, Phil Lind Chair in US Politics and Representation at the University of British Columbia, writes on regulatory politics and administration, the presidency, Congress, interest groups, elections, public opinion, and public policy. His book, *Politics of Deregulation* (co-authored with Martha Derthick), was the 2003 winner of the Aaron Wildavsky Award of the American Political Science Association.

MARTHA RHODES ROBERTS conducts research and public service for the University of Florida Institute of Food and Agricultural Science. Former Deputy Commissioner of Agriculture, Florida Department of Agriculture and Consumer

Services, she was Chairman of the Suwannee River Partnership. She holds a doctoral degree in microbiology from the University of Georgia.

EBERHARD ROEDER is Environmental Health Program Consultant with the Florida Department of Health, Bureau of Onsite Sewage Programs. His research has centered on contaminant hydrology, utilizing computer simulation models to understand groundwater pollution. He received a doctorate in environmental systems engineering from Clemson University, and was previously assistant in geological sciences at Florida State University, where he studied aquifer storage.

LAWRENCE S. ROTHENBERG is Max McGraw Distinguished Professor of Management and the Environment, and Co-Director of the Ford Center on Global Citizenship at the Kellogg School of Management, Northwestern University. He has written extensively on the political and economic determinants of policy choices. His most recent book is *Environmental Choices: Policy Responses to Green Demands*.

B. SUZI RUHL is director of the Public Health and Law Program for the Environmental Law Institute. A member of the board of the Center for Health, Environment and Justice, she served on Florida's Environmental Equity and Justice Commission, and as Vice Chair of the Community Environmental Health Advisory Board for the Florida Department of Health. Ruhl has a J.D. and an M.P.H from the University of Alabama, Birmingham.

PAUL A. SABATIER is professor of Environmental Science and Policy at the University of California, Davis. He has published in policy, management, and environmental journals on the role of scientific information in the public policy process, and other issues in environmental policy and implementation.

JOHN T. SCHOLZ is Frances Eppes Professor of Political Science at Florida State University. His research analyzes the problems of developing and maintaining cooperative solutions to collective action problems, emphasizing the role of policy networks, private partnerships, and collaborative government programs in resolving collective problems involved in resource management.

MELLINI SLOAN studies urban and regional planning at Florida State University, with research interests in the evolution and utility of demand side management of water resources. Former staff of the Florida Department of Environmental Protection, she has a master's degree in environmental engineering sciences from the University of Florida.

BRUCE STIFTEL is professor of Urban and Regional Planning at Florida State University and writes on collaborative processes in environmental decision making. A graduate of the University of North Carolina at Chapel Hill, Stiftel was editor of the *Journal of Planning Education and Research*, and president of the Association of Collegiate Schools of Planning.

LAWRENCE SUSSKIND is Ford Professor of Urban and Environmental Planning at Massachusetts Institute of Technology, where he heads the Environmental Policy Group. Founder of the Consensus Building Institute, a not-for-profit provider of mediation services around the world, Susskind has pioneered the development of environmental dispute resolution techniques and is currently involved in consensus building efforts in Israel, Canada, Japan, Brazil, Mexico, and Holland as well as collaborative efforts with the U.S. Geological Survey and the EPA's Office of Environmental Justice.

INTRODUCTION

The Challenges of Adaptive Governance

John T. Scholz and Bruce Stiftel

GROWTH AND DEVELOPMENT in twenty-first century America impose increasing stress on natural systems, which in turn increases conflicts among the diverse users of these systems. Problems among water users are evident even in states like Florida with abundant rainfall: declining water levels and saltwater intrusion in wells, polluted and diminished flows in springs and rivers, collapse of sinkholes, dried marshes, disappearing lakes, red tides, and habitat destruction, and other problems documented in the case studies in this volume. Scientific study of natural ecosystems is increasing our awareness and understanding of these problems and of possible solutions, but even advanced ecological knowledge has limited power to help resolve conflicts among competing demands for resources.

Problems on this scale inevitably involve collective action challenges of great complexity. Current governance structures are often the product of successful attempts to resolve collective action problems among users of a single type (e.g., urban, industrial, agricultural). This volume focuses on adaptive governance, which we define as a new generation of governance institutions for resolving collective action problems that occur between different types of resource users. Water quality, water supply, and habitat conservation are often the province of specialized authorities at local, state, and federal levels that support successful exploitation of water resources. Ironically, the very success of these specialized agencies brings about the expanding range of water conflicts that we study—conflicts that emerge as decisions by one authority impact other authorities and the users they govern. We call such conflicts second-order collective action conflicts, in contrast to the often simpler first-order conflicts that the specialized authorities were created to manage.

These second-order conflicts tend to be geographically defined by an integrated natural system—a river, an aquifer, a natural watershed—governed by multiple agencies. The evolving decision processes often utilize collaborative techniques from the field of conflict resolution. Such innovative processes have multiplied in recent decades (Bardach 1998) and are now a significant presence in natural resource governance (Wondolleck and Yaffee 2000). Our goal is to understand the challenges facing these innovative processes, and how they can help current institutions adapt to emerging second-order collective action problems.

Our concern with adaptive governance parallels adaptive management in developing techniques to deal with scientific uncertainties, as we discuss later, but extends them to include uncertainties about human institutions as well as the natural system. Sustainability requires that the two be compatible, and at times it may be more prudent to experiment with human than natural systems. Adaptive governance also aims to resolve conflicts among competing users in a manner that enhances joint gains while minimizing negotiation costs, but emphasizes that the resolution of conflict in the human system is valuable only if it leads to sustainable use of the natural system.

New Water Conflicts and the Second-Order Collective Action Problem

For most of the twentieth century, federal governance of water was dominated by development-oriented projects to increase the efficient use of water for specified purposes—projects that all congressional representatives sought for their own districts (Milbrath and Goel 1977; McConnell 1966). The U.S. Army Corps of Engineers and the U.S. Bureau of Reclamation developed the expertise and political base to design and build large projects that exploited apparently unused water resources to provide agricultural irrigation, hydroelectric power, and navigation (Reisner 1986). In areas of over-abundant water, canals were built to drain lands for agriculture and provide navigation (Blake 1980).

During the last quarter of the twentieth century, this development orientation came into increasing conflict with health and environmental concerns, which spawned new specialized agencies including the U.S. Environmental Protection Agency (EPA) and state counterparts like Florida's Department of Environmental Protection. Governance issues shifted to means of ensuring water quality with the Federal Water Pollution Control Act (U.S.C. §1251 et seq.): Congress began large-scale financing to help municipalities upgrade public waste treatment facilities, while EPA rules determined the kinds of investments required by private dischargers into surface waters (Rothenberg 2002; Clark and Cantor 1997). In addition, environmental concerns about the impact of water quality on endangered species brought new agencies like the U.S. Fish and Wildlife Service and state counterparts into water governance.

This shift to environmental issues mirrored the shift in political power away from agriculture and toward the cities and then to the suburbs, where environmental quality and the recreational use of water are valued as much as its economic development (Rothenberg 2002). In Florida, the state we feature in our case studies (see map, page 12), the longtime dominance of politics by agriculture and the northern panhandle counties gave way to the rapidly growing cities in the south. The combined shift in issues and demographics continues to play out in the water conflicts that we study, complicating the use of adaptive governance to resolve complex governance issues (Colburn and DeHaven-Smith 1999; DeGrove 1984; Pelham 1979).

Conflicts that arise under conditions of stress on the natural system frequently arise because of unexpected responses from the natural system, and require the affected agencies to deal with unfamiliar issues beyond their established expertise. They frequently involve new stakeholders, unfamiliar with the agencies' procedures, who challenge existing policies and established procedures. Conflicts among governing authorities create ambiguity about the proper arena for resolving the conflicts and solving the underlying problems. Coordinating policies across fragmented arenas could produce considerable benefits for stakeholders jointly affected by the decision of specialized authorities, but this requires combinations of expertise, authority, and representation of users that are not yet an established part of the institutional structure governing water resources.

Florida's Water Conflicts: the Case Studies

Of course, conflict is nothing new to observers of U.S. water policy. Water policy in the arid western states was in open conflict a century before the riparian east was forced to think in such stark terms. Today, however, there is no quarter of the United States in which water remains an abundant resource with effectively infinite assimilative capacity and the ability to readily satisfy the needs of all user groups. Lines of battle abound; many points of view compete for our understanding of who owns water, who should have the right to use it, what technologies are appropriate to employ, and what user risks are acceptable (Young 2005; Blomquist, Schlager and Heikkila 2004; Dzurik 2002; Ingram 1990; Maass and Anderson 1978).

Nowhere in the eastern U.S. is water more the subject of controversy than in Florida. Exuberant population growth and economic development, coupled with fragile groundwater stocks and delicate ecosystems, make Florida a center of tensions between the development industry and the environmental movement. It was in Florida that some of the earliest battles between development and preservation interests were played out: in the 1960s, proposals for an Everglades jetport and for a Cross-Florida barge canal led to national changes in environmental politics (Purdum 2002; Fernald and Purdum 1998).

Despite the battles over water rights, Florida's water policy has a history of innovation. It was in southern Florida that many of the technological challenges of canalization and wetland dredge-and-fill were first overcome. The state has the nation's most aggressive desalinization program and the most ambitious plan for developing aquifer storage and recovery systems. Greywater reuse, sprayfield disposal, and deep well injection were pioneered here. The state's regional water management districts have provided a national model for resource basin-based decisionmaking. Some of the nation's earliest and most notable multiparty collaborations in water policy were accomplished here, providing early examples of successful adaptive governance.

In short, institutional responses to Florida water conflicts illustrate both the challenge of political fragmentation of authority and the promise of pragmatic

American problem solving. The eight Florida case studies assembled in this volume offer a wealth of empirical richness about the second-order collective action problems confronting communities in all states. By selecting only Florida cases, we can provide essential details of the framework of conflict while sparing readers the considerable institutional complexity across 50 different state environments. Although the study context resembles the eastern more than the arid western states, we believe that the adaptive governance challenges analyzed here are relevant throughout the nation and across all natural resource issues. We hope that Florida's position at the leading edge of eastern states water conflict will more than compensate for our exclusion of the better-known first-order western water conflicts.

Each case examines conflicts in which multiple authorities confront novel situations that challenge existing procedures and authorities. Throughout this volume we will refer to cases by italicizing the place name from the table of contents. Two cases involve disputes over water quality: *Fenholloway* illustrates the expansion of conflicts over discharges permitted by private companies, and *Suwannee* illustrates the agricultural community's responses to increasing requirements to control previously unregulated non-point source pollution. Two of the cases, *Tampa Bay* and *East Central Florida*, focus on conflicts over water supply among competing municipal users. And finally, four cases explore conflicts involving the intricately connected aspects of water systems and habitat restoration. *Ocklawaha* explores the issues of removing a dam whose original purpose has disappeared. *Apalachicola* documents the tri-state dispute over allocations of water for different users and habitat preservation in the Apalachicola-Chattahoochee-Flint river system. *Aquifer Storage*, the exception to our place name reference because it applies across the state, explores efforts to expand use of a relatively new technology for underground storage of water. *Everglades* considers the Everglades restoration, the granddaddy of all Florida water conflicts involving almost all stakeholders and water issues in the entire region of South Florida.

The types of institutions and stakeholders vary across cases. Some conflicts bring the state legislature, administrative offices, and courts into novel situations. Most involve "advisory processes" with different levels of statutory guidance, ranging from ad hoc meetings to expanded interpretations of existing permit writing procedures and rulemaking authority to custom-designed statutory approval from the Florida Legislature and the United States Congress. Some decision processes attempt to forestall future conflicts, future regulatory expansion, or future challenges to permit decisions. Some institutional innovations appear to provide stable resolution of underlying conflicts, while others have failed and have perhaps exacerbated conflicts.

Adaptive Governance

As second-order conflicts have grown over the past decades, some specialized resource managers have adopted the tools of adaptive management (Gunderson et al. 1995; Lee 1993). Adaptive management focuses on the problem of managing resources when faced with inadequate knowledge and uncertainty about the natural system—we refer to this below as the challenge of scientific learning. A set of analytic and administrative tools let managers utilize the experimental techniques of science to test policy hypotheses during implementation. Policy then adapts to new knowledge about the natural system. Adaptive management techniques are particularly useful where there is broad agreement on policy goals but not about the appropriate means to achieve these goals. Scientific experimentation may then settle disputes about the most effective means.

Adaptive management fits well with the popular view of democracy in which elected officials determine policy goals, and specialized authorities then resolve conflicts about means. But the new water conflicts generally belong to the class of "wicked problems" (Rittel and Webber 1973) that do not fit this view because goals and means are inseparable. The policy tradeoffs involve a complexity of both human and natural systems that strains the ability of legislatures to make meaningful decisions to guide the adaptive management process.

If adaptive management seems ill-suited to the resolution of second-order conflicts within the existing political framework, conflict resolution suggests an alternative basis for adaptive governance of natural resources: bring the critical users, experts, authorities, and organized interests together into specialized negotiating frameworks designed to elicit mutually advantageous agreements. The well-developed study of negotiations and settlements can tell us much about adaptive governance, particularly the design of discussion forums.

But resolving the conflict among well-organized interests and competing authorities is not all that is involved in a sustainable resolution of current and predictable future conflicts. In order to be relevant to adaptive governance, conflict resolution must deal with limitations of both the human and natural systems. Solutions must account for not just the likely responses but also surprises from the natural system—the primary concern of the adaptive management approach. They must earn support not only from participating interests and authorities, but also from affected users and the broader political system. To provide sustainable solutions in the face of uncertainties about both the human and natural systems, second-order conflicts may require fundamental changes to governing institutions, changes that enhance the adaptive capabilities of the system without destroying the capabilities of agencies to manage resources effectively within the limited scope of their authority.

Adaptive governance, then, involves the evolution of new governance institutions capable of generating long-term, sustainable policy solutions to wicked problems through coordinated efforts involving previously independent systems of users, knowledge, authorities, and organized interests.

The contributors to this volume believe that successful governance of water and other natural resources in the twenty-first century depends on our ability to create adaptive institutions, and that this ability will depend on resolving five challenges to adaptive governance:

1. Representation (Who should be involved?)
2. Decision Process (How can authorities and involved stakeholders reach policy agreements that serve them well?)

3. Scientific Learning (How can policy makers develop and use knowledge effectively?)
4. Public Learning (How can resource users and the relevant public develop common understandings as a foundation for consensual policies and policy processes?)
5. Problem Responsiveness (How well do decisions achieve natural resource management goals, including sustainability, equity, and efficiency?)

1. Representation

The first challenge is to determine who should be represented in the new procedures and institutions, with what resources, and with what authority. Conflict arises from the harms imposed on some users by others. To resolve conflicts within the political system, the harmed users must identify the source of harm, articulate their interests, and create political alliances able to change existing rules and authorities (Easton 1965). But identification, articulation, and aggregation of interests are all subject to the collective action problems involved in getting individuals to work together (Olson 1965), which may leave the political landscape dominated by organized economic interests. When different types of affected users remain unorganized or ineffective participants, should adaptive governance institutions proactively expand representation?

Why is adaptive governance concerned about representation? In the democratic ideal, citizens have equal access to decision processes intended to change public policies, although this ideal is seldom realized. Pragmatically, greater representation may help obtain needed resources and reduce challenges from represented interests (Flyvbjerg 1999; Moe 1980; Wengert 1971). The standard of representation in localized, ad hoc decision processes must be acceptable to the greater political community to avoid being overturned by administrative, legislative, or judicial institutions (Meyer 2001). Furthermore, broader representation may transform issues and make broader, mutually advantageous tradeoffs possible (Susskind 1981; Pateman 1970).

On the other hand, broader representation also complicates the negotiation process. Processes with more parties may require more formality, share fewer common perspectives, develop less personal trust, and increase the time required for deliberations, all of which may undermine the critical belief by participants that the process will be fruitful, particularly if some of the included groups are clearly hostile to the process.

2. Process Design

The second challenge is to develop decision mechanisms that satisfy the groups involved in the process. Effective representation requires a reasonable understanding of what the represented group prefers over a likely range of policy outcomes, an ability to translate these preferences into policy options, and resources and skills to gain approval of preferred policies. When representatives have restricted access to information, technical knowledge, and decisionmaking, participation can be simply a façade used to claim legitimacy. Purely advisory deliberations may have little impact on final decisions made elsewhere (Harter 1992; Arnstein 1969).

Full voting rights do not guarantee a working process. Too often, the normal venues for developing policy options operate without sharing information. Too often, they do not facilitate innovative problem solving. Too often, solutions do not endure. Through the 1980s, experiments with "win–win" process designs led to proposals for new participatory frameworks that emphasized cooperation over competition. These experiments, including negotiated rulemaking, policy dialogues, and various mediated forms, became more widespread in the 1990s, but vexing questions remain (Forester 1999; Innes and Booher 1999). How can scientific and technological knowledge be made accessible to unsophisticated interests, and how can technical creativity overcome partisan agendas? How do these new participatory structures interact with traditional decisionmaking authorities and political leadership? Should government agencies—frequently the primary participants in collaborative institutions—have a different status than water users?

What role should leaders, administrative authorities, facilitators, and public interest groups play in developing representation for unorganized interests (Jones et al. 1992; Shaw 1988)? What qualifications should be required of paid experts, public interest groups, or self-appointed representatives who undertake such work? Are they sufficiently knowledgeable about the issues to develop positions, defend them against opposition, and incorporate the detail necessary to make them viable within broader policy decisions? Since these are particularly important in the ambiguous and unstructured settings in which the new water disputes arise (Rubin and Sander 1988), can safeguards be developed to protect the legal rights of stakeholders in nontraditional proceedings?

Process design also affects the accountability of representatives to the group they represent, and hence the legitimacy of decisions approved by representatives. For example, the transparency of fully public proceedings may ease the concern that representatives may sell out their constituency. Of course, public negotiations in partisan settings may also unduly constrain the delicate processes of compromise and creativity. In addition, transparency is elusive for many complex technical issues of water governance; neither constituents, nor their local, state, or federal elected representatives, nor the media have the time, expertise, and inclination to observe and interpret the technical discussions among experts. Even if they did, the scientific and technical uncertainties inherent in the majority of policy decisions could make constituents doubt the sincerity of their representatives. This is particularly true for adaptive governance techniques in which decisions are expected to change in response to new knowledge. Since unanticipated consequences and the resultant policy changes generally affect the distribution of costs and benefits, constituents may question how well they were represented in the initial decision. Promoting trust and accountability between representatives and their constituency while allowing the flexibility required for developing innovative, adaptive policies (Bianco 1994) remains one of the most critical and least understood challenges of process design.

3. Scientific Learning

Science is critical for much of environmental policy. Water policy questions are often steeped in hydrology, geology, and more recently in a wide range of life sciences. Yet the policy process is not sophisticated in its understanding or use of science. Sometimes policymakers and stakeholders expect science to speak with a single voice, and to have answers readily available. Often they expect a higher degree of certainty than scientists can offer, particularly when available data do not allow the specificity that policymakers require (Adler et al. 2001; Nelkin 1992; Ozawa 1991). Often policymakers are asked to decide questions when the scientific community would prefer to assemble more data. This leads to appeals for delay, to expenditure of scarce resources for data collection that might otherwise be deployed in program implementation, and to decisions that do not enjoy the full confidence of interests.

Policy processes that complement the norms of scientific progress and knowledge are more likely to use science effectively. At least three dimensions are important. First, different specialists reflect different perspectives on the human and natural systems. Decision venues that clarify and contrast differences may lead to productive syntheses, but may also exacerbate conflicts. Second, policy processes can advance scientific knowledge by providing a forum for experts to review existing results and design research projects to fill gaps. Finally, policy decisions may be structured as scientific experiments in which outcomes are monitored to test the critical assumptions on which the policy rests, as envisioned in the adaptive management literature.

These concerns raise many questions about the role of science in adaptive deliberations. What kinds of information are used in the decision process, and from what sources? How are scientific conflicts resolved, including disagreements over data, methods of analysis, and underlying theories? How should decisionmakers and stakeholder representatives be involved in determining which issues to resolve and how to resolve them? Is science biased: who are the scientists? Who pays them? Are "independent" scientists active in policy-related research? How influential is scientific analysis, and how is it integrated into the process? Does a technical focus on "science" exclude other valuable knowledge from resource users?

4. Public Learning

Policy deliberations often involve a relatively small number of elites, frequently representatives from associations and organized interests that seek to represent resource users or other stakeholders (Ostrom et al. 1994). Implementation of agreements, however, will generally require the cooperation of users themselves and the resources and legitimacy from political institutions and a broader section of the public. To what degree can we expect adaptive governance to lay the groundwork for such cooperation through the process we refer to as public learning?

Public learning implies that users learn about the broader consequences of their actions, the reasons for restricting particularly harmful actions, and the available alternative policies that reduce harms and enhance benefits. It is equally important that they accept the legitimacy of the decision process itself and the lessons learned by stakeholders about conflicting demands on water resources and how to resolve them.

The general public and the primary political institutions (legislative, executive, judicial) are involved in public learning as well. They will control what resources of authority and funding can be used to implement the agreement, and what alternatives are available to disgruntled stakeholders who wish to challenge the agreement. The political institutions and the public segments they respond to ultimately determine the long-run legitimacy of these decision processes and the policy decisions they make. Education of stakeholders reshapes the nature of the conflict; education of the greater public redefines the context in which the conflicts are played out.

Public learning also includes the process by which individuals and society learn to value new outcomes and the decision processes that can bring them about. This type of public learning centers on how resource users, experts, authorities, and organized interests can explore and change—some would say transform—their understanding of the legitimacy of opposing interests and the effects of their own behaviors on the resource stocks and flows. This is an old concern of political theory, eloquently described by J.S. Mill in the nineteenth century as "social development" (e.g., Mill 1962). But it is also a key concern in recent planning theory (Hillier 2002; Healey 1997; Friedmann 1987). For example, Meyer (2001) argues that local communities are increasingly likely to organize to oppose federal restrictions imposed to protect endangered species, in part because of different perceptions about the resilience of nature and the rights of property owners and local communities to exploit local resources. Long-term resolution of conflicts—particularly those involving habitat conservation—relies in part on the transformation of beliefs about the natural system and about the legitimacy of restrictions and the process that produces them.

How can decision processes foster exploration of beliefs and preferences by user groups and the general public, and at what cost? The authors in this volume will explore the characteristics of arenas for public discussion that transform the perceptions and preferences of the public and reshape the underlying conflicts. How does new information emerge and affect public debate and policy decisions? Does the process lead to surprises for stakeholders and the general public? Do these surprises change beliefs and preferences? How is the press involved? How can representatives transform the preferences of their constituency rather than just reflect unconsidered preferences and beliefs?

5. Problem Responsiveness

Decision processes are judged by how well the policy decisions and implementation respond to the underlying source of conflict. The policy sciences have long assessed public policy in terms of economic efficiency and equity, and these remain central to our view of whether any given water policy deliberation or institutional change is successful. Sustainability has recently become a central test of the environmental consequences of policy.

Efficiency is traditionally viewed as the promotion of greatest benefits to the largest portion of the public, a goal that defies simple calculation. Pareto's widely-known early formulation quickly proved to have limited applicability, yet no alternative decision rules are universally accepted. Cost–benefit analysis is often used as a surrogate for efficiency, but has been widely criticized for incompleteness (Ackerman and Heinzerling 2004). As a result, there will often be competing views of the efficiency outcome for any given policy. Still, the concept is useful for thinking through a policy's long-term economic and social impacts (Young 2005; Weimer and Vining 1999, *331–378*).

Efficient outcomes may have differential effects: there are often winners and losers. So assessments of efficiency must consider equity. Here, too, there is no simple answer. Society may be divided into many different deserving sub-groups. Justice can be interpreted using a variety of criteria. Still, we recognize that natural resource policies have to accommodate differences in wealth and income, geographic distribution, race, ethnicity, and gender, while simultaneously recognizing the legitimate interests of rival water user communities (Sager 2002; Mueller 1976).

Both efficiency and equity face the difficult tradeoff between stability and adaptability. Stability allows full exploitation of the potential gains made possible by existing institutions. To the extent that first-order collective action problems can be reasonably resolved, stability allows for the discovery of efficient investments and behaviors. This stability is particularly important for large-scale, long-term investments. Adaptability, on the other hand, sacrifices some of the localized efficiency of stable first-order institutions in order to encourage exploration of greater global efficiencies. Policy tools and institutional changes intended to enhance adaptability need to be judged in terms of the costs to short-term efficiency as well as the potential long-term gains. For example, local consensus-based negotiations may improve scientific and public learning, but may also impose considerable costs due to the extended uncertainties imposed by lengthy negotiations. These costs are probably worthwhile for resolving second-order conflicts that by definition create the potential for greater global efficiencies. But the costs may be less justifiable for resolving the chronic conflicts about the distribution of benefits within first-order institutions when second-order conflicts are rather minor.

Finally, we have come to expect that environmental policy will be sustainable, that the resource uses agreed upon will remain viable beyond the foreseeable future. The origins of sustainability can be found in early twentieth-century notions of conservation and sustained yield, but in the years since the 1987 Bruntland Commission report (World Commission 1987), traditional stewardship of natural resources has been extended to recognize the need for socially stable outcomes (Krizek and Power 1996; Caulfield 1978).

Conclusions

In sum, we believe that the five challenges are key to creating successful second-order institutions capable of adaptive governance. Effective representation requires innovative means of ensuring legitimate, meaningful involvement of affected groups in a manner congruent with the norms of representation in the overall political system. Deliberative process design requires a decision process capable of recognizing and articulating the needs of affected users and consolidating them into practical policies that can achieve consensus. Scientific learning requires incorporation of diverse scientific viewpoints to answer critical policy questions, monitor outcomes, and challenge policy assumptions inconsistent with new findings. Public learning requires transformation of beliefs and preferences about legitimate water rights and decision processes among users as well as the community and political institutions. Finally, problem responsiveness requires that the process produces an efficient, equitable, and sustainable solution.

The case studies and analyses in this volume draw preliminary lessons about the institutional characteristics of adaptive governance. We do not attempt to determine the "ideal" solutions for adaptive governance of water resources, particularly since characteristics that help to meet one challenge may exacerbate another. Adaptive water governance above all else requires attention to the specifics of any conflict. Our goal is the more modest one of contributing to the ongoing debate a better understanding of the tradeoffs involved in different institutional designs and their ability to tackle the challenges of effective representation, deliberative process design, scientific and public learning, and problem responsiveness.

Case Study Locations

Water Management Districts
NWFWMD: Northwest Florida Water Management District
SFWMD: South Florida Water Management District
SJRWMD: St. Johns River Water Management District
SRWMD: Suwannee River Water Management District
SWFWMD: Southwest Florida Water Management District

Notes:
* Aquifer Storage and Recovery involves locations throughout the state

Part I
Case Studies of Water Conflicts

CHAPTER 1

Florida's Water Management Framework

Richard Hamann

TO UNDERSTAND THE case studies in this volume, one needs a map of the institutions responsible for management of Florida's water resources. This chapter provides the minimum description necessary to participate in the debate over institutional change.

Water management in Florida, as in most states, is conducted by a variety of institutions at every level of government. Local governments have comprehensive authority, but limited resources and jurisdiction. Water management districts are the most significant of a series of regional agencies, with authority over every aspect of water management. At the state level, the Florida Department of Environmental Protection (FDEP) supervises the water management districts, regulates most sources of water pollution, manages sandy beaches, regulates the quality of public water supplies, and administers most state lands, including the beds of navigable waters. Other agencies with significant responsibility include the Department of Community Affairs (DCA), which administers state programs for comprehensive planning and land use regulation, the Florida Department of Health (FDOH), which regulates septic tanks, and the Florida Department of Agriculture and Consumer Services (FDACS), with jurisdiction over agricultural sources of water pollution and state forests. The state's fish and wildlife are managed by the Florida Fish and Wildlife Conservation Commission. Federal agencies participate in the management of Florida water resources as land managers, regulators of water pollution sources, funders of water-related projects, and regulators of activities adversely affecting fish and wildlife. The relationships among these agencies and levels of governance are complex.

Local Governments

Local governments have extensive authority over activities that can affect water resources. They have primary responsibility for regulating land use. Through the adoption and implementation of a local comprehensive plan, they regulate the effects of growth on the environment, and provide for the concurrent delivery of such services as water supply, wastewater treatment, and stormwater control (*Fla. Stat.* Ch. 163, Pt. II (2003)). Some local governments, especially in urban areas, have sophisticated programs for regulating water pollution, petroleum storage facilities, hazardous waste facilities, and other potential sources of water pollution. These programs are based on either a delegation of the state's pollution control authority or the local government's own authority.

In many cases certain governmental services are provided outside the normal structure of general purpose local governments. Port authorities, for example, may levy taxes and operate ports and navigation channels. Extensive mosquito control operations, involving water management and the application of pesticides to wetlands, may be conducted by mosquito control districts. Water control districts build and operate works for drainage and flood control in much of Florida. Landowners, who pay assessments for the drainage, have voting rights in proportion to their landholdings. As new communities are developed in Florida, much of the responsibility for providing services and infrastructure is being given to community development districts established and controlled by land developers.

Regional Agencies

Florida has several regional institutions that influence the management of water resources. Eleven regional planning councils are composed of local government representatives and gubernatorial appointees (*Fla. Stat.* 186.501-.513 (2003)). Established in 1975, the regional planning councils originally had stronger authority over local land use decisions. Although they are now mandated to develop Strategic Regional Policy Plans, they have little authority to implement them and primarily serve to assist and coordinate local government members on a voluntary basis.

A more significant institution for regional cooperation is a Regional Water Supply Authority (RWSA), (*Fla. Stat.* §373.1962 (2003)), which may be created by interlocal agreement among local governments.[1] Regional water supply authorities can develop and manage facilities for regional water supply. As described in *Tampa Bay,* the first was the West Coast Regional Water Supply Authority, which has been reinvented as Tampa Bay Water (*Fla. Stat.* 373.1963 (2003)). Several other RWSAs have since been created. By joining together in a RWSA, local governments can pool their resources to secure permits, raise capital, and resolve interlocal disputes over water supplies.

The most important regional institutions for managing water resources are Florida's five water management districts (*Fla. Stat.* Ch. 373 (2003)). These have the most comprehensive authority, the greatest technical expertise, and the largest financial resources of any state or regional agencies dealing with water in Florida. The water management districts were organized along surface water hydrologic boundaries, not along political lines. They operate independent of local governments and with a great deal of autonomy from state control. Water management districts are significant actors in nearly all the case studies in this volume.

The water management districts were based on a model developed in South Florida beginning in the 1920s. Efforts to drain the Everglades and provide flood control to the east coast had been based on the creation of special taxing districts. Major flooding after a tropical storm in 1947 overwhelmed the system, and the state requested assistance from the U.S. Congress in constructing a massive expansion. For Congress to authorize a project by the U.S. Army Corps of Engineers, a local sponsor would be needed with the ability to purchase necessary lands and operate the system. The Florida Legislature created the Central and Southern Florida Flood Control District to perform those duties, and gave it the assets of the earlier districts and authority to levy *ad valorem* taxes. Based on that federal–state partnership, a massive system of levies, impoundments, canals, pumps, and discharge structures was created.

It worked well, and after hurricane Donna inundated southwestern Florida in 1960, a similar project was developed for that area. The Southwest Florida Water Management District (SWFWMD) was created to serve as the local sponsor. SWFWMD was given additional regulatory authority over groundwater withdrawals.

A severe drought in 1970–1971 led to a reexamination of the entire framework for land and water management and the passage of several landmark pieces of legislation. One of those bills, the Florida Water Resources Act of 1972, extended the system of water management districts to the entire state and gave them a uniform set of powers and duties (Maloney et al. 1972). Voters subsequently approved a constitutional amendment extending the authority to levy ad valorem taxes to the remainder of the state.[2]

Each of the water management districts has a governing board of gubernatorial appointees, which hires an executive director, adopts a budget, issues administrative orders, and adopts rules. To coordinate water policy at the state level, the districts are subject to the general supervisory authority of FDEP, their budgets must be approved by the governor, governing board appointments must be confirmed by the Senate, and the rules and orders of the districts may be appealed to the governor and cabinet.

The authority given to the water management districts was broad to begin with, and has been significantly expanded. They can build and operate canals, dikes, pumping stations, and other water management structures for drainage or water supply purposes. They can also buy and manage land. They can conduct scientific and technical studies, and are required to formulate several kinds of water management and water supply plans, including plans for the restoration of degraded surface waters and the supply of water in areas of predicted shortage. They are required to provide technical assistance to local governments, which are in turn required to consider district plans in developing their own water supply plans. As part of the planning process, the districts are required to adopt minimum flows and levels (MFLs) for streams, lakes, and groundwater. MFLs must protect the water resources of the area from significant harm. In addition, the districts can reserve water from consumptive use for the protection of fish and wildlife and public health and safety. Although the requirement to adopt minimum flows and levels was part of the original 1972 legislation, the districts have done so only recently in response to litigation by citizens.

Florida's common law for resolving water use conflicts, as in the other eastern states, was based on riparian rights. All of the owners of riparian land had a right to make reasonable use of the resource. Unlike the prior appropriation system developed in the west, there was no well-defined property right vis-à-vis subsequent users; all users were protected, not just those who withdrew water from natural systems. Now the allocation of water for consumptive use is an administrative function of the water management districts. Most consumptive use is subject to regulation to prevent harm to the water resources. With limited exemptions, there are no rights to use water except as permitted by the districts. Permit applicants must demonstrate that a proposed use will not interfere with existing legal users and is "reasonable-beneficial," and consistent with the public interest. Permits are subject to renewal after a fixed duration, with the possibility of competing applications. Applicants have the right to a twenty-year permit during which period they are protected against interference and exempt from compliance with new conditions. The use of water may be further restricted, however, during droughts under the terms of water shortage plans or pursuant to water shortage orders. The transport of water outside the basin of origin, across county boundaries, or even from one district to another may be authorized, but there is a preference for local sources and the development of alternative water supplies.

Environmental resource permitting (ERP) is the other major regulatory program of the districts.[3] As part of an effort in the 1990s to reduce regulatory overlap and duplication, implementation of this program is shared with the FDEP.[4] ERP regulates drainage, stormwater systems, dredge and fill activities, and other structures to manage water quality, downstream flooding, groundwater levels, wetland functions, and habitat for aquatic and wetland-dependent wildlife. It considers measures to avoid or minimize impacts, and requires the remaining impacts to be offset through compensatory mitigation. ERP may authorize the use of navigable waters held in trust for the public and construction seaward of coastal construction control lines along Florida's sandy beaches.

State Institutions

Florida's state institutions are divided into the familiar legislative, executive, and judicial branches. The Legislature plays a strong role in Florida water management and believes that it is the only institution with the authority to establish water policy. In 1981, the FDEP adopted a "state water policy," which it had developed in consultation with the water management districts (*Fla. Admin. Code* r. 62-40). In 1997, the legislature renamed it the "water resource implementation rule" and required that it be submitted to the legislature before it could go into effect. The Legislature has also severely constrained the authority of the districts and other agencies to adopt rules unless they are very specifically authorized by statute, and has imposed other limitations on the ability of state agencies and local governments to protect water. The Harris Act (*Fla Stat.* 70.001 (2003)), which requires compensation for regulations that "inordinately burden" private property, is a notable example.

Florida's executive structure is somewhat unusual. The Governor can appoint some agency heads, but there are also several independently elected officials with authority over their own agencies. The Commissioner of Agriculture, for example, controls the Department of Agriculture and Consumer Services (DACS), which has significant responsibility for controlling water pollution from agricultural activities and the use of pesticides. DACS has a primary role in implementation of the *Suwannee Partnership*. DACS also manages state forests. In water policy debates, DACS can be an influential advocate for agricultural interests and rural landowners. The Attorney General has taken a strong and independent stand in pursuing water-related litigation, for example to ensure the protection of navigable waters. For some purposes, the Governor and members of the Cabinet serve as members of collegial bodies with shared governing responsibility. For example, the Trustees of the Internal Improvement Trust Fund, composed of the Governor and Cabinet, are vested with the ownership of most state lands, including the beds of navigable waters, and thus are responsible for protecting the public trust in those waters. Decisions of the water management districts may be appealed to the Land and Water Adjudicatory Commission. Certain land use decisions are made by the Administration Commission. Meetings of the Governor and Cabinet consist of a series of meetings of such commissions and boards. The number of cabinet officers has been reduced,[5] and more executive powers have been delegated to the Governor, but many important executive decisions related to water are made collegially.

The constitution also establishes the Florida Fish and Wildlife Conservation Commission, an independent executive agency with exclusive control over the fish and wildlife resources of the state. Its members are appointed by the Governor, but that is the extent of his authority. The Public Service Commission (PSC) is an independent agency created by the Legislature; it sets the rates charged by utilities for water supply and wastewater treatment and their service areas. The decisions of the PSC can have profound effects on how these services are provided.

The heads of the other state agencies with responsibility for water management are appointed by the Governor. The most significant are FDOH, FDCA, and FDEP.

FDOH regulates septic tanks and other onsite treatment and disposal systems that can pollute water. It also regulates private wells to ensure drinking water safety, as well as the safety of public swimming areas.

FDCA acts for the state in land use planning and regulation. Certain large-scale developments known as Developments of Regional Impact (DRI) have been subject to review by FDCA and the regional planning councils (*Fla. Stat.* §380.06 (2003)). This program has been seriously eroded, however, and is of declining importance. FDCA also implements a program for designated Areas of Critical State Concern (*Fla. Stat.* §380.05 (2003)). Under this program, specific environmentally significant and vulnerable areas of the state may be designated and "principles for guiding development" to protect their resources may be adopted that are binding on local governments. The protection of water quality, wetlands, and other water resources has been one of the principle reasons for designating such Areas of Critical State Concern as the Florida Keys, Green Swamp, Big Cypress Swamp, and Apalachicola Bay. Local governments have been successfully pressured into more adequately managing growth in these areas, but the Legislature has limited the program to only 5 percent of the state's land area.

The most pervasive land use planning and regulatory program is the Local Government Comprehensive Planning and Land Development Act (*Fla. Stat.* Chap. 163, Pt. II (2003)). FDCA is responsible for ensuring that local governments comply with minimum standards and criteria for the development of comprehensive plans (*Fla. Admin. Code* r. 9J-5). Many of these standards are directly relevant to the protection of water quality and quantity, fish and wildlife, and other water resources. One of the criticisms of the comprehensive planning process has been the failure to adequately link it to water management planning (Angelo 2001). Local governments are now required to consider water management district plans, and to develop their own water supply plans that meet all projected needs for a ten-year period.

FDEP has the most comprehensive responsibility for water management of any Florida state agency (*Fla. Stat.* Ch. 161, 253, 258, 259, 373, 376, 378, 403 (2003)). It was created in 1993 by merging the Department of Natural Resources (DNR), an agency that had been accountable to the Governor and Cabinet, with the Department of Environmental Regulation (FDER), accountable to the Governor. This merger was touted as a way to reduce conflict and overlap in environmental programs. The secretary of FDEP is appointed by the Governor. The responsibilities of the agency range from acquiring and managing state lands to controlling air and water pollution. To the extent that pollution control programs are delegated by federal agencies, FDEP is generally the recipient of that authority and funding.

The land management responsibilities of FDEP are extensive. Title to the beds of navigable waters is vested in the Trustees, but FDEP provides the staff to administer those lands. Part of the navigable waters is designated as Aquatic Preserves, with special management by FDEP. In addition, FDEP administers numerous state parks, preserves, reserves, greenways, trails, and recreation areas, many of which preserve or provide access to the state's water resources. All of the state's land acquisition programs are administered by FDEP, and it oversees Everglades restoration programs. It also regulates the reclamation of mined lands.

The regulation of sources of water pollution is one of the primary responsibilities of FDEP. FDEP adopts water quality standards for both surface and groundwater and is in the process of adopting total maximum daily loads for hundreds of impaired water bodies. It regulates sources of water pollution, including industrial facilities, electrical power generators, underground injection wells, landfills, petroleum storage and transport facilities, mining operations, certain stormwater systems, and facilities for the storage or disposal of hazardous wastes. For the activities that FDEP otherwise regulates, it also issues Environmental Resources Permits. FDEP administers the state programs for cleaning up sites where hazardous wastes, toxic substances, petroleum products, and dry cleaning solvents have contaminated the soil and water. FDEP supervises the water management districts and adopts the water resource implementation rule. It also implements the state's coastal zone management program, regulates development along Florida's sandy beaches, and conducts an extensive program of beach nourishment. Much of the information on which water management decisions are based comes from FDEP's Florida Geological Survey. FDEP regulates the quality of potable water supplied by larger systems.

All states are required under the federal Clean Water Act to adopt water quality standards establishing minimum levels for water quality. They consist of two parts, a use classification and both narrative and numeric criteria for that class of waters. There are five classes:

- Class I, potable water supplies;
- Class II, shellfish propagation or harvesting;
- Class III, recreation and propagation and maintenance of a healthy, well-balanced population of fish and wildlife;
- Class IV, agricultural water supplies; and
- Class V, navigation, utility, and industrial use.

Each class specifies narrative and numeric quality criteria. Unless otherwise designated, all surface waters in the state are Class III. Only the Fenholloway River was ever Class V and it has been reclassified to Class III. In addition to these classifications, waters in a protected area or with exceptional recreational or ecological significance may be designated as an Outstanding Florida Water (OFW) or an Outstanding National Resource Water (ONRW). These waters are given additional protection against significant degradation. Where surface waters do not meet water quality standards, the state is required to establish a Total Maximum Daily Load (TMDL) and take action to achieve those standards through reductions in both point and nonpoint sources of pollutants. One of the reasons for development of the *Suwannee Partnership* has been the desire of agricultural sources of water pollution to avoid being regulated and to avoid designation of the Suwannee as an "impaired water," which would require a TMDL.

The standards for groundwater are very different. Groundwater is classified according to its potability, the level of total dissolved solids, and the degree of confinement. Existing or potential sources of drinking water are given the highest levels of protection. In addition, Maximum Contaminant Levels (MCL) for drinking water, issued by the federal Environmental Protection Agency, also serve as minimum criteria for groundwater quality.

Federal Agencies

A variety of federal agencies influence the management of water resources in Florida. These include the Environmental Protection Agency (EPA) and the Departments of Interior, Agriculture, Commerce, and Defense. They regulate directly or through delegation to the state, fund state activities, provide technical expertise to other agencies, and manage extensive areas of land and water.

The Department of Defense might seem an unlikely water manager, but it has, in fact, been one of the most influential. There are several major military installations in Florida, and the management of those lands is of critical environmental importance. The most important Defense entity, however, has been the U.S. Army Corps of Engineers (USACE), in both construction and regulation. The Corps of Engineers and its local partners play key roles in several of the case studies. The USACE raised Hoover Dike around Lake Okeechobee in 1930, constructed the Central and Southern Florida Flood Control Project in the 1960s and 1970s, and is now working to implement the Comprehensive Everglades Restoration Plan and to complete restoration of the upper St. Johns River. In the 1960s it began a series of similar projects in southwest Florida that would have channeled, impounded, and diverted rivers throughout the region. USACE dammed the *Ocklawaha* River as it began to construct the Cross Florida Barge Canal. The intracoastal waterway, numerous navigation inlets, and beach nourishment projects throughout Florida have been constructed by the Corps.

The regulatory side of USACE has been equally active. In the 1960s the Corps began to assert jurisdiction over construction of waterfront property that involved dredging and filling mangrove wetlands and shallow waters. Much of the coast would have been destroyed without that intercession. The jurisdiction of USACE was extended farther inland in 1972 and the Corps, working with other federal agencies, has been responsible for limiting the destruction of wetlands in the Everglades and many other parts of Florida.

Most federal regulation of water pollution has been delegated to the EPA. EPA may work in partnership with other federal or state agencies. The process of review, oversight, and coordination can be complex and time-consuming. EPA works with the USACE to ensure that wetland permits protect the resource. EPA sets the criteria for evaluating permits, reviews applications, and can veto permits issued by USACE. EPA establishes the minimum criteria for the state to use in setting water quality standards and regulating many dischargers of pollutants. EPA also can exercise direct permitting authority. EPA sets the minimum criteria for drinking water and groundwater quality. Although most sources of pollution are regulated by FDEP, it is often implementing a federal program and thus subject to oversight by the EPA. The use of pesticides and other toxins is regulated by the EPA. The remediation of sites that have been contaminated by hazardous wastes or other toxic substances may be compelled, carried out, or funded by the EPA. The EPA is an important source of scientific research on aquatic ecosystems, water quality, environmental health, and technologies for pollution prevention and control.

The Department of Interior (DOI) also has significant water-related activities in Florida, operating principally through the National Park Service (NPS), the U.S. Fish and Wildlife Service (FWS), and the U.S. Geological Survey (USGS). NPS manages such important water-dependent areas as Biscayne and Everglades National Parks and the Big Cypress National Preserve. Much of the conflict in South Florida over water management has centered on potential impacts to these areas. FWS also manages significant natural areas that are dependent on water management, but unlike NPS, it is also responsible for fish and wildlife on private lands. Before a federal agency like the Corps of Engineers implements a project or issues a permit affecting water resources it must consult with FWS to determine how to avoid, minimize, or mitigate the impact on fish and wildlife. The duty becomes even stronger if the project may affect species that are listed as endangered or threatened. Even nonfederal projects on private lands may be limited to avoid harm to listed species. Because so many endangered or threatened species are vulnerable to changes in the flow or quality of water, to construction in surface waters or wetlands, or to other activities that occur in the water such as navigation, FWS takes part in many decisions affecting water resources. The development of adequate science to fulfill these responsibilities and provide technical assistance is a related function of the agency. Developing good science is the principal function of USGS. Water management in Florida depends on the geologic, hydrologic, and biological research conducted by USGS scientists over many decades. One final Interior subdivision, the Bureau of Indian Affairs, supervises two tribes in Florida, the Miccosukee and the Seminole. As dependent sovereigns, these tribes have been authorized by Congress to operate outside the control of the state for many purposes and may manage land use, fish and wildlife, water quality, and consumptive use of water within their trust territories.

The Department of Agriculture is almost as multifaceted as DOI. It has similar land management responsibilities, with three major national forests protecting large watersheds and aquifer recharge areas. The restoration of degraded watersheds, including the *Ocklawaha River*, has been a primary objective of the U.S. Forest Service. The Department of Agriculture has a more extensive agenda, however. Its price support program for sugar helps to subsidize the degradation of water resources. On the other hand, programs administered by the Natural Resources Conservation Service are helping to restore wetlands and water quality and to preserve agricultural and silvicultural land uses.

Finally, there is the Department of Commerce, which houses the National Oceanic and Atmospheric Administration (NOAA). Marine and coastal areas are a particular concern of this agency. Marine mammals and endangered and threatened species that live in marine environments are protected by the National Marine Fisheries Service (NMFS), an agency within NOAA. There is also an independent Marine Mammal Protection Commission. The regulation of fisheries in federal waters is governed by NMFS. The National Ocean Service administers an extensive network of marine sanctuaries and reserves. The waters surrounding the Florida Keys, for example, are part of the Florida Keys National Marine Sanctuary, with an active program of regulation, education, research, and enforcement.

Conclusion

The regulatory map of agencies and levels of government involved in the management of Florida water is fragmented and overlapping. Conflict can slow decisionmaking and impede implementation of even the best plans. Accountability suffers when responsibility is widely shared. Centralization and consolidation, however, have their own problems. The scale of water management issues is both geographically and substantively vast. As the *Apalachicola* case study demonstrates, the quality and quantity of water resources in Florida is directly affected by activities in much of Georgia and Alabama. If the effects of air pollution and climate change are considered, the scale is global. Political consolidation is thus most difficult. Furthermore, water is affected by virtually everything we do as human societies: agriculture, transportation, manufacturing, urban development, and recreation. No single agency can ever have all the authority needed to fully protect and manage water resources. With consolidation unlikely, linkages, alliances, and other arrangements that work across geographic and institutional boundaries are of heightened importance.

Notes

1. The authority of local governments to create regional institutions by interlocal agreement is not limited to water supply (*Fla. Stat.* 163.01 (2003)).

2. The existing districts in southern and southwest Florida already had the power to tax; the amendment was enacted because of the support of those populations. A compromise with powerful legislators that allowed it to be placed on the ballot limited the authority to .05 mills in the Panhandle, as compared to 1.0 mills in the remainder of the state (Carter 1974).

3. The Northwest Florida Water Management District has limited funds because of the constitutional limitation on millage, and the legislature has refused to fund a state program, so the ERP program has not been implemented in that rapidly developing area.

4. Either FDEP or a WMD has jurisdiction over an ERP application, depending on the kind of regulated activity. The criteria are the same no matter which agency issues the permit. ERPs for the most common kinds of activities are issued by water management districts.

5. The Cabinet now consists of the Governor, the Attorney General, the Chief Financial Officer, and the Commissioner of Agriculture (Fla. Const. Art IV, §4(a)).

CHAPTER 2

Suwannee River Partnership

Representation Instead Of Regulation

Ayşın Dedekorkut

THE SUWANNEE RIVER PARTNERSHIP began in 1998 as a government-led voluntary effort by agricultural producers and conservation groups to avert a water quality crisis through incentive-based reduction of nutrient discharges. With key stakeholders reluctant to participate and their scope of authority constrained by legislation, the water managers designed a process that integrated scientific knowledge and won a high level of voluntary participation by large agricultural producers. The case illustrates the difficulty of obtaining full representation in consensual ad hoc policymaking, but shows the value of collaborative scientific fact finding, and may serve as a model for promoting public learning.

The Middle Suwannee River Basin in North Central Florida includes Lafayette and Suwannee Counties (Figure 2-1) and is a major recharge area to the Floridan Aquifer, the main source of water supply in the Suwannee River Basin. Because of its water quality and scenic nature, the Suwannee River was designated an Outstanding Florida Water[1] in 1979. The predominantly rural, agricultural basin is characterized by highly permeable limestone overlain by transmissive sandy soils and numerous sinkholes, and the aquifer is extremely susceptible to nonpoint source pollution (FDEP 2003). The basin is home to hundreds of residential and commercial septic systems, about 300 row crop and vegetable farms, 44 dairies, and about 150 poultry farms (Woods 2001). Without nutrient management, waste from these agricultural operations pollutes the Suwannee River and the Floridan Aquifer.

The nitrate form of nitrogen from non-point sources is the main pollutant in the basin. Elevated nitrate concentrations could significantly damage the ecology of the river and the Suwannee River Estuary, principally through eutrophication, which causes algae blooms, depletion of oxygen, and fish kills (Katz et al. 1999). Nitrate concentrations higher than 10 mg/L in drinking water constitute a health hazard to children and pregnant women.

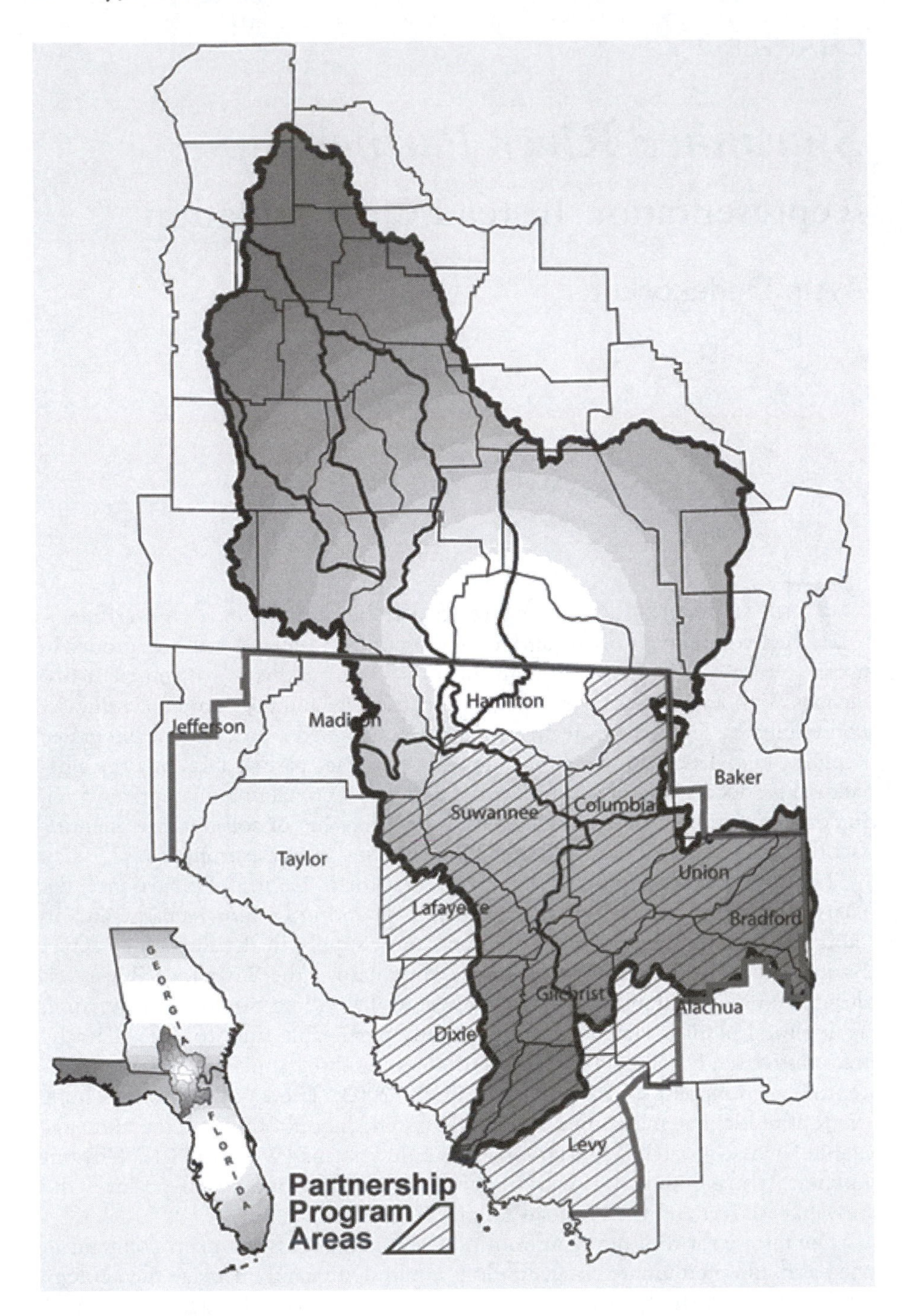

Figure 2-1. *Suwannee River Basin*

The Partnership programs affect the Middle Suwannee and Santa Fe Basins, which together constitute the potential source of 60–70 percent of the nitrate load to the Suwannee River Estuary (Suwannee River Partnership 2002). They primarily involve changing agricultural practices and better management of human waste. The three major targets of the Partnership are dairy, poultry, and row crop farmers.

History of the Partnership

Nitrogen readings in the Middle Suwannee River Basin have doubled over the last 20 years (Ritchie 2002). Groundwater from this watershed is affecting the surface water quality of the Suwannee River via springs and seeps in the riverbed (Hornsby et al. 2002b). Springs in the Middle Suwannee River Watershed have nitrate-nitrogen concentrations ranging from 1.2 to 17 mg/L (Hornsby et al. 2002a), and in 1990 nitrate concentrations in water from the Upper Floridan aquifer in parts of Suwannee and Lafayette Counties exceeded the maximum contaminant level of 10 mg/L set by the Environmental Protection Agency (EPA) for drinking water (Katz and Hornsby 1998). In water year[2] 2001, the Suwannee River Basin accounted for 98 percent of the 3,067 tons of nitrate-nitrogen and 78 percent of the 909 tons of total phosphorus that were transported to the Gulf of Mexico by the six area rivers. The Middle Suwannee River Basin, which covers only 8.6 percent of the total Suwannee Basin, accounted for 45.5 percent of the annual nitrate-nitrogen load delivered to the Gulf by the Suwannee River, whereas the Santa Fe River covering 5.7 percent contributed 15.8 percent (Hornsby et al. 2002a).

The Santa Fe and Lower Suwannee Rivers are Class III water bodies, designated for recreation, propagation, and maintenance of a healthy, well-balanced population of fish and wildlife. The Suwannee River Estuary is Class II, designated for shellfish propagation or harvesting. The Florida Department of Environmental Protection (FDEP) has determined that the Middle and Lower Suwannee River, the Lower Santa Fe River, and the Suwannee River Estuary may not be meeting their designated uses based on legal criteria (§ 62-303, Fla. Admin. Code) and are potentially impaired by excessive nutrients and algal mats (Suwannee River Partnership 2002). These rivers and the Suwannee River Estuary were on Florida's 1998 303(d) List of Impaired Surface Waters, and are on the 2002 Update (Suwannee River Partnership 2002). Thus FDEP is required by federal law to reduce nitrate levels in the River through regulatory measures under the Total Maximum Daily Loads[3] (TMDL) process. These measures can limit how farmers may fertilize or irrigate their crops, restrict how livestock producers manage the wastes from their animals, or change how septic systems are designed, installed, and maintained (IFAS 2002).

The Formation and the Structure of the Partnership

Suwannee River Water Management District's (SRWMD) spring and groundwater monitoring results, which showed elevated nitrate levels in the Middle

Suwannee River Basin, created widespread concern. SRWMD research showed the following estimated sources of nitrogen in Suwannee and Lafayette Counties: fertilizer 45 percent; poultry 33 percent; dairy cows 10 percent; beef cows 5 percent; atmospheric 6 percent; and people 1 percent (Hornsby and Mattson 1998). In the early 1990s FDEP conducted several Hydrologic Unit Area Demonstration Projects with EPA grants to assess dairy and poultry contributions to nonpoint source water pollution in the Middle Suwannee River Basin and to demonstrate new technology in the treatment of animal waste. U.S. Department of Agriculture's Natural Resources Conservation Service (NRCS) prepared plans to retrofit area dairy and poultry farmers with the best available technology and management practices for animal waste management, utilization, and disposal (FDEP n.d. a).

In 1991, at the request of the Lafayette and Suwannee River Soil and Water Conservation Districts, NRCS initiated a land treatment project in the Middle Suwannee River Area under the authority of Public Law (PL) 566, the Watershed Protection and Flood Prevention Act. In 1993, the NRCS approved PL-566 provisions for financial and technical assistance to dairy and poultry farmers to install conservation measures under Long Term Contracts (LTCs). The Small Watershed Program for the Middle Suwannee was the first PL-566 project in the nation authorized for the sole purpose of groundwater quality protection. In 1995 the SRWMD provided supplemental cost share funds to dairy farmers through the Surface Water Improvement and Management (SWIM) program to expedite development and implementation of LTCs (Swicegood 2001). In 1996, the NRCS and SRWMD agreed to a cost-share arrangement to help dairy farmers in Suwannee and Lafayette counties to implement best management practices (BMPs) to improve animal waste management (SRWMD n.d. b). In 1999, the SRWMD approved funding of a special technical team to expedite development and implementation of LTCs for poultry farmers. In addition, the Florida Department of Agriculture and Consumer Services (FDACS) offered poultry farmers additional cost-share funds to install their LTCs (Swicegood 2001).

By 1998 a number of state and regional agencies were collecting data on the sources and amount of nitrogen in the river, conducting research to understand the ecosystem better, and investigating funding opportunities to help farmers implement BMPs. In response to the need for coordination between numerous regulatory agencies that had authority over aspects of the issue, SRWMD requested during the Governor's Water Resources Coordinating Council meetings in December 1997 that the districts and other agencies coordinate their resources and research to develop a watershed strategy to control nitrates (Matthews and Grippa 1997).

The PL-566 Small Watershed Program for the Middle Suwannee and the Hydrologic Unit Area Demonstration Projects paved the way for the Partnership. The parties created the Suwannee River Partnership, chaired by FDACS, by signing the Agreement in Principle on Nutrient Management in the Suwannee River Basin on January 25, 1999. This agreement showed the signatories' intent to cooperate, but was not legally binding.

The 24 members that signed the original agreement include the EPA, the U.S. Geological Survey Water Resources Division, and USDA NRCS from the federal government; FDEP, the Department of Community Affairs, the Florida Department of Health (FDOH), and FDACS from the state government; SRWMD as a regional government agency; Lafayette and Suwannee Counties as local governments; Lafayette Soil and Water Conservation District, Suwannee River Soil and Water Conservation District, and Suwannee River Resource Conservation and Development Council as conservation groups; as well as industrial, agricultural, and research groups. The mission of the Partnership is to determine the sources of nutrient loads to the basin, and to work with local land users to minimize future nutrient loading through voluntary, incentive-based programs for protecting the environment and public health. The group has concentrated initially on the Middle Suwannee River Basin (Agreement 1999). Five Technical Working Groups with designated responsible agencies were created to address priority nutrient sources: Animal Waste Management and Fertilizer Management Groups—FDACS; Human Waste Group—FDOH; Monitoring Group—FDEP; and Outreach and Education Group—Florida Farm Bureau (SRWMD 2003a). Each group drew up plans to reduce nutrient loading in the water resources of the basin. To coordinate the work of all the participants, the Partnership created a position of full-time coordinator funded jointly by SRWMD, FDEP, and FDACS. While physically located at the Live Oak headquarters of the SRWMD, the coordinator, Darrell Smith, is an employee of the Partnership.

Components of the Partnership Programs

The Partnership relies on voluntary cooperation and financial incentives for implementation of BMPs to reduce nitrogen loadings in the basin. BMPs are economically and technologically feasible changes in farming practices designed to reduce polluted runoff and conserve water. Because nitrates come from a variety of sources such as animal waste, human waste, and fertilizers, the plan proposes BMPs for each possible source. In poultry, BMPs relate to litter storage, dead bird disposal, and nutrient management. On dairies they relate to waste, and on row crops BMPs include irrigation and nutrient management (Loop 2001).

Dairy, poultry, and row crop farmers receive assistance to improve animal waste management and fertilization techniques from the USDA-NRCS PL-566 Small Watershed Program for the Middle Suwannee,[4] the Environmental Quality Incentive Program (EQIP) administered by USDA Farm Service Agency, and the NRCS, along with state funds from the SRWMD and FDACS cost share programs (Florida-Agriculture.com 1999). Each farmer requesting assistance through the NRCS will have a Dairy Comprehensive Nutrient Management Plan or a Poultry Conservation Plan that meets NRCS specifications and practices. Table 2-1 breaks out the costs of BMP implementation, and Table 2-2 shows the amount of money committed by each group to the Partnership programs.

The BMP Quality Assurance Program verifies that best management practices are maintained properly over a long period and provides assistance to farmers in resolving problems with BMPs. FDACS inspectors verify that BMPs are being maintained through routine on-site evaluations. When deficiencies are found, the two technicians of Lafayette and Suwannee Counties Soil and Water Conservation

Table 2-1. *Cost-Share Funding for BMP Implementation*

Type of operation	*Average cost of BMP implementation per operation*	*Cost-share funding sources*
Dairy	$254,000	65% (up to $100,000): NRCS, 50% of remaining (up to 50,000): SRWMD SWIM funds Remaining: Farmers
Poultry	$62,500	65% (up to $100,000): NRCS 50% of remaining (up to $9,240): FDACS Remaining: Farmers May also qualify for USDA EQIP
Row crop	Not yet determined as implementation began recently. Varies significantly based on farm size	Was 75%, reduced to 50% with the new Farm Bill (up to $50,000): USDA EQIP FDACS may pay 25% (up to $50,000 if the farmer agrees on certain practices) Remaining: Farmers

Table 2-2. *Funding Sources and Needs for BMP Implementation*

Funding sources and needs	*Funds for dairies*	*Funds for poultry farms*	*Funds for row crop farms*
USDA NRCS PL-566 funds	$3.7 million	$4.2 million	None
USDA EQIP funds	None	$500.000	$1.5 million (2001 and 2002)
SRWMD SWIM cost-share funds	$1.65 million	None	None
FDACS cost-share funds	None	$832.000	$188,300 (FY 2001–2002)
Farmers' contribution	$4 million	$1.4 million	Not yet determined
Funding Needs	$2.7 million		$5.7 million

Districts funded by the Partnership follow up with farmers to correct problems (Suwannee River Partnership 2002). Fifteen poultry farms and four dairies were checked in 2001, and in 2002, 30 poultry farms and 4 dairies. The On-Farm Research Program includes BMP Effectiveness Demonstration Projects, or "319 projects,"[5] to verify that BMPs are effective in reducing nitrates lost to groundwater and surface waters. The Water Assessment Regional Network (WARN) monitors groundwater and surface water to identify trends in water quality over time (SRWMD n.d. a). The Florida Farm Bureau Federation certifies farmers participating in the Partnership programs through the County Alliance for Responsible Environmental Stewardship (CARES) Program. The Partnership recognized 37 CARES participants in 2001 and 43 participants in 2002 (Crawford 2002).

Santa Fe Basin Expansion

The success of the programs in the Middle Suwannee Basin prompted the Suwannee River Partnership to expand into the Santa Fe River Basin in 2002 by hiring another jointly funded coordinator to draft a plan of action to begin Partnership programs. The priority area in the Santa Fe Basin is the Lower Santa Fe Basin, which includes the reach of the river scheduled for development of a TMDL standard in 2007. The suspected sources of nitrogen in this area differ from the Middle Suwannee Basin, with a notable absence of poultry and a higher percentage from fertilizers and human sources.

The staff's recommendation for the Santa Fe Basin includes using the existing Steering Committee structure to provide direction for the initiative and using presentations to recruit additional partners in the basin. Possible interest groups include Soil and Water Conservation Districts of the three counties in the priority area, industry groups, and civic and special interest groups. The recommendation includes providing a benefit package to the agriculture industry in the basin. For reducing the nitrates originating from agricultural practices, techniques similar to those in the Middle Suwannee Basin will be used. However, comprehensive research is necessary to deal with the human waste problem by reinstating the Human Waste Technical Working Group with the addition of local stakeholders in the basin (H. Thomas 2003). This was not a big problem in the Middle Suwannee Basin, consequently the working group did not have to seek new technologies and practices.

Only producers in the Middle Suwannee River area from Dowling Park to Branford are eligible for the current PL-566 program. An extension of this to the Santa Fe Basin does not appear likely; there are limited funds for each state in this program, and Santa Fe has to compete with proposals from the rest of the state. The Partnership is currently looking for alternative funding sources.

Court Challenges to FDEP's Implementation of the Clean Water Act

Environmentalists have not joined the Partnership because they view it as a loophole for farmers to avoid pollution limits, and have opted to challenge the FDEP's participation in court. FDEP program administrator Daryll Joyner denies that the state has created a loophole because under EPA guidelines, those who propose exempting waterways from TMDLs must explain how much pollution will be reduced and by what date. As a result, FDEP requires "serious" documentation of the steps taken to reduce pollution. Farmers also deny that they are taking advantage of a loophole; they argue that the voluntary program encourages participation in pollution reduction more than regulations (Ritchie 2002).

Critics of the nonregulatory approach of the Partnership focus on two main points: whether the farmers would maintain the BMPs in the long run in a voluntary program, and whether BMPs are effective. BMP Quality Assurance Program and BMP Effectiveness Demonstrations address these points. Partnership coordinator Darrell Smith (2003a) believes these two programs set Suwannee River Partnership apart from other nonregulatory approaches. Carol Kemper (2003) of EPA Region 4 agrees, and claims that the Partnership is different because it takes care of Quality Assurance.

Nevertheless, the proposed Impaired Waters Rule (IWR) of FDEP, which allows the Partnership to provide an alternative to TMDL regulations, was challenged at the state level by six public interest groups and two individuals who filed a petition to have it overturned as an invalid exercise of delegated legislative authority under §120.56, Fla. Stat. The petition alleged that nearly every provision of the proposed rule failed to comply with the Florida Administrative Procedure Act and was thus invalid. On May 13, 2002, Department of Administrative

Hearings (DOAH) Administrative Law Judge Stuart Lerner issued a 368-page final order in DOAH Case No. 01-1332RP which concluded that the proposed rule was a valid exercise of delegated legislative authority.[6] Florida's First District Court of Appeals ruled in FDEP's favor in the appeal of the final order on May 20, 2003, upholding Judge Lerner's order (Borkowski 2003). The IWR has survived the state court challenges and has been effective since June 10, 2002.

Some of the same parties who filed the state challenge also challenged the EPA, claiming that EPA has failed to review the Florida IWR as a change in water quality standards, as required by the Clean Water Act (CWA). FDEP claims that the IWR is not a water quality standard or a change to existing water quality standards, so the agency moved to intervene and was allowed to file as an intervening defendant. FDEP asserted that the plaintiffs' argument relied on a misinterpretation of Florida water quality standards and that the alleged change was non-existent. The Court ruled in favor of EPA and FDEP on May 29, 2003 and assessed costs against the plaintiffs (Borkowski 2003). The Eleventh Circuit Court of Appeals reversed the order on October 4, 2004 and remanded the case back to the district court to determine the practical effect of the IWR on state water quality standards. As of March 2005 the case is on remand to the U.S. District Court for the Northern District of Florida (Stephens 2005).

In a separate action, environmental group Save Our Suwannee and supporters sued FDEP for violating state law and the CWA by failing to require Concentrated Animal Feeding Operations (CAFOs) to apply for National Pollution Discharge Elimination System (NPDES) permits.[7] While in voluntary partnership with FDEP to reduce pollution from their operations, the CAFOs are not required to obtain any groundwater or surface water permits. At trial, FDEP justified its partnership program under §403.0611, Fla.Stat., which allows the FDEP to "explore alternatives to traditional methods of regulatory permitting, provided that such alternative methods will not allow a material increase in pollution emissions or discharges." On March 5, 2004, the presiding judge ordered FDEP to immediately require all CAFOs to apply for NPDES permits or demonstrate the applicability of an exemption. FDEP's appeal to the Court of Appeals was denied on March 2, 2005 (Stephens 2005).

A third challenge involves petitions asking EPA to withdraw Florida's NPDES permitting authority. The plaintiffs allege that Florida has failed to administer the program in accordance with the CWA. Their justification includes FDEP's failure to require NPDES permits for CAFOs and its use of the IWR to change Florida's water quality standards.

The case is currently pending (Stephens 2005b). The impact of these court challenges on the Partnership, particularly the CAFO decision requiring some Partnership participants to file for permits, remains uncertain at this point.

Representation

The Partnership was formed under the leadership of the SRWMD and FDACS as an ad hoc process that brought together parties that historically did not collaborate—regulator and regulated—to decide collectively on how to deal with the problem. The Suwannee River Partnership and the Middle Suwannee Basin Work Plan are built on the concepts of partnership and coordination, and stakeholder involvement is critical. Participation is encouraged through public meetings within the basin, meetings among partners, and periodic progress reports (SRWMD 2003a). Three of the four major groups interested in water quality of the Suwannee River Basin—agriculture, regulatory agencies, and scientists—are involved in the Suwannee River Partnership. Environmental groups are not despite encouragement from the Partnership (Roberts 2003b), primarily because they fear that the Partnership will undercut regulation (Webster 2003), as evidenced by their court challenges to FDEP decisions.

To ensure farmer participation, Partnership representatives attended meetings of farm organizations to promote the program. They sought out the most respected farmers early on and explained the program to them, as Lubell discusses in Chapater 18. When these were convinced and talked about the benefits, other farmers became interested (Webster 2003). Farmers participated in the Technical Working Groups as well. According to Glenn Horvath (2003) of SRWMD, farmers signed up once they saw that BMPs could save money, but without financial incentives they might not have been able to participate. Under the TMDL approach farmers would receive less funding and fewer would use BMPs.

Some individuals were also crucial to the success of the Partnership: Congressman Allen Boyd, Representative Dwight Stansel, Senator Richard Mitchell, and Commissioner of Agriculture Charles Bronson have been helpful in getting cost-share funds from federal and state governments (Smith 2002). Many groups praised the leadership of SRWMD Executive Director Jerry Scarborough and Deputy Commissioner of Agriculture Martha Roberts.

Design of the Decision Process

Legislation has delineated available policy choices for dealing with nutrient management in the Suwannee River Basin. EPA's TMDL requirement with the CWA proposes one alternative and FDEP's IWR provides another. The local bodies choose between these two policies. FDEP may have provided an alternative to EPA requirements in reaction to typical rules made by the federal government that usually are not flexible enough to fit the needs of specific areas—an example of a natural resource question not being addressed at appropriate level of government.

The main executive agencies involved in the Partnership are FDEP and FDACS, with the support of Lafayette and Suwannee Counties' local governments and other groups. EPA oversees and can decide whether there is sufficient assurance of water quality improvement to make TMDLs unnecessary. The Partnership Steering Committee informed EPA representatives during the Steering Committee Meeting on April 16, 2003 in Tallahassee that they did not feel like partners with EPA and that their programs were never acknowledged by the agency as valid, despite the active participation of groups not covered by any regulatory agency or statute (Roberts 2003a). EPA promised to increase management participation in the Partnership and to keep contributing funds. EPA representative

Curry Jones (2003) emphasized that the voluntary incentive-based approach was the best way to deal with this kind of a problem, and that EPA intended to use Suwannee as a model for other areas within EPA Region 4, especially in northern Georgia. However, during the subsequent review of the Reasonable Assurance Documentation, EPA voiced concerns about the comparatively low participation of row crop farms (82 out of about 300 have plans). The Partnership claims this is mostly due to lack of funding, not intent; 207 row crop farms have signed up with the program, some of which are waiting for funding and technical assistance. EPA's concerns were prompted by the anticipated lawsuit by a Florida environmental organization. With national attention on the Suwannee River, they wanted to make sure that their case was watertight when they reviewed and accepted FDEP's impaired waters list and Reasonable Assurance Documentation for exempted waters (Jones 2003). Consequently EPA did not accept FDEP's proposal of keeping parts of the Suwannee River off the impaired waters list, but rather added these parts as low priority.

Scientific Learning

Environmental interest groups are not convinced that BMPs are effective means of improving water quality. The On-Farm Research Program of the Partnership includes a BMP Effectiveness Demonstration Project at one dairy, one poultry, and one row crop farm to measure pre- and post-BMP water quality. The Partnership will monitor wells at all three sites over the next five years to determine the effect of BMPs (Smith 2003b).

Still, the only significant scientific uncertainty is how soon water quality will actually improve. Even if the BMPs are working it will take a long time to observe changes in the groundwater. As a result, achieving water quality levels for Outstanding Waters designation is not projected to occur before 2028.

Public Learning

The Steering Committee responsible for coordinating the Middle Suwannee Basin Work Plan, composed of the chairs of each Technical Working Group, meets monthly. Progress in research and implementation is reported in these meetings, which are open to public (SRWMD 2003a). The reports, and summaries of these meetings, are shared through an e-mail list and the Partnership website. The Partnership views communication as essential to the success of the Basin Work Plan. As a result, a Technical Working Group dedicated solely to outreach and education, chaired by the Florida Farm Bureau, works to increase understanding and support among stakeholders and the general public (SRWMD 2003a).

The Partnership has numerous research and education programs:

- University of Florida's Institute of Food and Agricultural Sciences (IFAS) Manure Lab in Live Oak provides manure analysis and application rate recommendations to growers in the Basin, and is funded by several members of the Partnership. The Manure Lab Committee, composed of IFAS faculty, FDACS, and Partnership staff, meets monthly to consider education programs related to animal waste management in the Basin, and publishes a newsletter for producers (Smith 2003b).

- The Mobile Irrigation Laboratory (MIL), administered by the Suwannee River Resource Conservation and Development Council and funded by FDACS, evaluates irrigation systems for efficiency, makes water conservation recommendations to farmers, and demonstrates the benefits of water conservation for both nutrient management and water quality. The MIL team evaluated around 75 systems in 2002 (Smith 2003b).
- NRCS and the Conservation Districts provide technical assistance for farmers to implement BMPs. In addition, FDACS and FDEP provide BMP follow-up assistance to farmers through the BMP Quality Assurance Program (Smith 2003b). Soil and Water Conservation District technicians advise producers regarding available BMPs, provide technical assistance, and convey feedback from the grower community to the Partnership agencies on technical assistance, cost sharing, research needs, and the success of voluntary efforts (SRWMD 2003a). Farmers view technicians as a valuable resource because they can talk to farmers in their own language and earn the farmers' trust (Barnes 2003). The technicians help farmers see the need for continuous management and better practices. Horvath (2003) maintains that the change will be accomplished through education, not the stick approach of regulatory programs.

Problem Responsiveness

The Suwannee River Partnership is one of the first basin-wide voluntary participation conservation programs in the nation (FloridAgriculture Viewpoint 2000). Glasgow (1999) maintains that through the voluntary efforts of producers and the technical and funding support of conservation partners, an impending water quality crisis has been averted without imposing mandatory rules and regulations. Early results show that farmers have kept more than 77 tons of nitrates from dairy wastes and 475 tons of nitrates from poultry wastes out of the aquifer through voluntary, incentive-based nutrient management practices.

Farmers participate in such a program for many reasons. While financial incentives are very important, some of the groups supposedly participated because they feared that regulation changes were "around the corner" (Horvath 2003). Farmers believed that the regulatory approach would force all farms to adopt the same standards regardless of their effectiveness with individual farms, whereas the Partnership approach makes individual recommendations according to farm-by-farm variations in the types of products, hydrological characteristics such as existence of sinkholes, etc. FDEP (n.d. b) praises the Suwannee River Partnership for working with farmers to develop customized plans rather than mandating a single regulatory program for all agricultural interests in the region.

According to Assistant Director of Agricultural Policy at the Florida Farm Bureau Federation Frankie Hall (2003), the flexibility and freedom allowed for by the voluntary approach is not only more efficient, but also appeals to the independent nature of the farmers. Such personal values influence reactions to voluntary versus regulatory programs.

Partnership participants argue that TMDLs would probably not achieve better practices and water quality because "people would be intimidated" (Joyner 2003), while with voluntary approaches, farmers are doing more than what they would do otherwise. Many of the BMPs they apply are not required under regulatory programs (Webster 2003). In addition, the proponents of the Partnership maintain that traditional regulatory programs like CAFO rules generally apply only to larger operations. In the Middle Suwannee Basin this includes only four dairy and ten poultry operations; in the Santa Fe Expansion one dairy in the priority area, and three dairies in the extended area would be treated as a point source and have to apply for a NPDES permit or demonstrate that they have no potential discharge (Seibold 2003). For FDEP it is harder to enforce regulations on many smaller farms. The Partnership is trying to cover operations not covered by other regulatory programs. As a result, BMPs are applied more extensively through the voluntary approach.

In addition to funding and fear of regulations, stewardship and trust play a role in farmers' participation in the program (see Chapter 18). As poultry farmer Nancy Barnes (2003) put it, "farmers can't be without soil and water." It is in their best interests to keep this resource healthy. Frankie Hall (2003) of the Florida Farm Bureau Federation says that building better relationships with the community is important to farmers and through this approach they "gain all around." The bad examples of the Dairy Rule imposed on the Lake Okeechobee watershed,[8] and TMDL regulations for the Everglades and resulting lawsuits, also played a role in making people see the value of the proactive approach. In one of the early meetings with farmers the Partnership brought a representative from the Everglades sugar industry who talked about the litigation and fighting between the agricultural interests, the regulatory agencies, and environmentalists, and asked the audience not to make the same mistake.

It is difficult to evaluate the fairness of the distribution of costs and benefits. Some believe that the Partnership is dominated by agricultural interests and is a way to funnel state and federal funds to farmers. In contrast to the cost-share funding that agricultural operations receive, industrial point sources are expected to cover full costs of water quality improvement measures. Poultry farmer Chuck Edwards (SRWMD 2003b) states that since keeping the river clean affects everybody that uses it, including the tourists, everybody is benefiting from the cost share, not only the farmers, and "society should pay for what they benefit from" (Edwards 2003). He further argues that everybody is responsible for the current state of the water, not only the farmers.

As of May 23, 2003, the Partnership had signed up 39 of the 40 dairies, 131 of the 139 poultry farms, and 207 of the 300 crop farms; 32 dairies, 126 poultry farms, and 82 crop farms in the basin have Management Plans and have started implementing Best Management Practices. The remainder are waiting for technical assistance or funding. The Partnership encourages farmers to sign up in advance for the program, because NRCS uses the long waiting lists to seek funding. The goal of the Middle Suwannee Basin Work Plan (Suwannee River Water Management District 2003) is to achieve 80 percent participation in BMPs for row crop and 100 percent participation for poultry and dairy farmers in the basin by 2008, when the PL-566 program that provides financial assistance to farmers will fully be implemented.

The Reasonable Assurance Documentation (Suwannee River Partnership 2002) cites significant progress toward the goal of clean waters. However, it is too early to forecast the outcome because the implementation of BMPs in all farms will not be completed until 2008, and restoring the Suwannee River Estuary to levels comparable to water quality at the time the rivers were designated as Outstanding Water Bodies in 1979 is to be accomplished by 2028 (Suwannee River Partnership 2002).

Conclusions

The Suwannee partnership has grappled with representation, but stands as a success story in process design and public learning. These successes appear to be due to a series of choices among which incentives, trust, voluntary cooperation, and mutual responsibility figure prominently.

The partnership began under threat of a legislative solution. Without such a threat only the farmers that were forced to do something would participate in partnership programs and they would do so without commitment. Moreover there were significant financial incentives; many doubt that the same number of farmers would have participated otherwise. The proposed reduction of EQIP cost-share funds from covering 75 percent of the costs of implementation to 50 percent is expected to reduce participation, and therefore the effectiveness of the program (Joiner 2003).

While farmers are well represented, environmentalists are not. The prevailing view in the environmental community has been that the Partnership opens a loophole for farmers to avoid pollution limits. This view led to legal challenges, some of which are currently pending.

The leadership and commitment of key people such as SRWMD Executive Director Jerry Scarborough and Deputy Commissioner of Agriculture Martha Roberts, as well as the support they were able to secure from people with political influence, such as Commissioner of Agriculture Charles Bronson and state and federal legislators, were crucial. This leadership may also have been critical in building trust among the Partnership participants, and that trust was certainly key to resolving the issues in a collaborative manner. The participation of the Farm Bureau Federation was key to building trust between the farmers and the regulatory agencies. Scarborough (2003) believes that the Farm Bureau gives credibility to the Water Management District in the eyes of the farmers.

Farmers showed good faith in participating in the program. However, because EPA rules on Impaired Waters do not allow delisting a water body before water quality improves, TMDLs have to be prepared for portions of Middle Suwannee and Santa Fe. This may cause the farmers who think they were promised no regulation to lose faith in the Partnership. The technical members attending a Steering Committee Meeting argued that the strategies used to improve water quality will be the same whether TMDLs are set or not, and this message needs to reach the farmers. Leadership was also sensitive to the need to recognize farmers for their efforts. Commissioner of Agriculture Charles Bronson personally presents the CARES signs to each farmer recognized through the program in a ceremony each year at Representative Stansel's farm. The expected technical support was also very important for the farmers.

The availability of scientific information and its wide distribution through extensive public education and farmer outreach programs aided agreement among the variety of interests. The bad examples of the Everglades and Lake Okeechobee demonstrated the likely outcomes of not collaborating, and participants knew they did not want a similarly lengthy and painful process. Significant doubts among farmers as to the efficacy of BMPs were assuaged by provisions for monitoring and re-evaluation.

Incomplete representation created legitimacy problems, but didn't prevent the parties from thoroughly reviewing the evidence and crafting workable solutions. Whether the absence of environmentalists will ultimately prove to be a serious liability depends on how successfully the Partnership can retain farmers.

Finally, the voluntary nature of the conservation programs is appealing to the farmers. As suggested by Florida Farm Bureau President Carl B. Loop Jr.: "voluntary Best Management Practices work. They work better than practices mandated through government regulation" (Crawford 2002).

Although participants advocate the Partnership as a successful alternative to regulation for addressing water quality problems, the series of lawsuits indicate that not everybody is convinced. The failure to involve environmental groups and provide adequate assurance about the effectiveness of Partnership activities remain the major weaknesses of the Partnership.

Notes

1. An Outstanding Florida Water is water designated worthy of special protection because of its natural attributes, and is intended to protect existing good water quality with stricter stormwater controls (FDEP 2003).

2. A "water year" is a 12-month period from October 1 through September 30, designated by the calendar year in which it ends. Thus, the year ending September 30, 1999 is called the "1999" water year (U.S. Geological Survey 2003).

3. A TMDL is a scientific determination of the maximum amount of a pollutant that a river, lake, or other surface water can tolerate without exceeding surface water standards that protect public health, wildlife, and habitat (FDEP 2001). EPA is responsible for setting TMDL standards for waters that were previously listed in 1998 even if they are not on the 2002 303(d) list unless water quality improvement takes place.

4. The Program is locally sponsored by the Suwannee and Lafayette Soil and Water Conservation Districts, and helps dairy and poultry farmers.

5. This project is usually referred to as the "319 project" because it was funded by EPA grants through Section 319 of CWA.

6. The final order can be found at http://www.dep.state.fl.us/water/tmdl/docs/IWRfinal-ruling.pdf.

7. EPA's CAFO rules require farming operations over a certain size to get a NPDES permit. In dairy, these are operations of 700 or more mature cows. In poultry, depending on the type of operation, operations over 30,000 (liquid manure operations), 82,000 (dry litter layer operations), and 125,000 (other dry litter operations) birds are considered to be CAFOs. Under the new Final Rule dated December 15, 2002, CAFOs must implement nutrient management plans that include appropriate best management practices to protect water quality. The deadline for FDEP to comply with this is April 2006.

8. § 62-670-500, Fla. Admin. Code, enacted by FDEP in 1987, required all dairies within the watershed and its tributaries to implement BMPs for reducing phosphorus flows into the lake (SFWMD 1997).

CHAPTER 3

Fenholloway River Evaluation Initiative

Collaborative Problem-Solving Within the Permit System

Simon A. Andrew

TO ATTRACT BUSINESS investment and revitalize the local economy after World War II, local officials in Perry, Florida approached the Florida Legislature with a request to reclassify the Fenholloway River for industrial purposes. In 1947, the request was granted (24952 *Fla. Stat.* 1338), and by 1954 Procter and Gamble Cellulose Company had built a pulp and paper mill in Perry. The Foley mill produces specialized kraft paper products and cellulose fibers, and is the primary employer in the county. For most of its half century of operation, this mill has been the subject of controversy.

The Foley mill uses chlorine, hypochlorite, sodium hydroxide, and chlorine dioxide to dissolve pulped logs and extract cellulose fibers. This process can discharge 50 million gallons per day of treated effluent, which constitutes an average of 85 percent of the Fenholloway River's flow. By 1992, this discharge had transformed what had been a local economic hero into a national environmental outrage. Amidst national debate regarding dioxin and looming legal challenges from environmental groups, Procter and Gamble sold its entire pulp business. The Foley Plant was sold to an investment business headed by former Procter and Gamble executives and the name changed to Buckeye Florida L.P.

This chapter traces the long path of public and scientific learning about the consequences of the discharge, which led to an intractable conflict over renewal of the required discharge permits. The story begins in a context of public opinion formed apart from the formal permitting system, and illustrates problems of fragmented authority and adversarial legalism that led to stalemate even within the relatively unified structure of the National Pollutant Discharge Elimination System (NPDES). Frustration with indecision finally brought about an innovative mechanism to expand the existing permitting process: the Fenholloway River Evaluation Initiative (the Initiative) in which the contending factions and primary authorities collaborated to evaluate policy options and agreed on a plan that would provide the basis for the contested permits. After relating the history of the case, we examine the Initiative from the perspectives of the five challenges of adaptive governance.

Foley and the Fenholloway: Growth of Concern

Public attention to environmental problems at the Foley plant grew slowly. In 1980, a group of scientists found that the Mosquito fish (*Gambusia holbrooki*) living downstream from the Foley mill had masculinized secondary sex characteristics (Howell et al. 1980), confirming suspicions that mill effluents were causing damage. This scientific finding was kept quiet, however, and received little media attention. A decade later, when Florida was hit by a severe drought in 1989–1990, residents near the plant complained to the local health authority that their well-drawn drinking water smelled like rotten eggs. Although state officials found no conclusive evidence of chemical contamination in June 1990 well-water tests, residents were advised not to drink local well water by the Florida Department of Health and Rehabilitative Services.

In 1991 the Tallahassee *Democrat* published a Pulitzer prize-nominated story on the Fenholloway (Hauserman 1991). The article stimulated debate, and environmental groups petitioned the Environmental Protection Agency (EPA) to challenge Florida environmental officials and upgrade the river from its industrial classification to a recreational river where "fish can survive and people can swim" (Multinational Monitor 1991).

Amidst the debate over water quality in the Fenholloway River, owners of a mobile home park filed suit in the Federal District Court in Tallahassee against Procter and Gamble later in 1991 for water contamination, which they argued had decreased the value of their property. And in 1992, about 100 local residents filed suit against the company for property damage, creation of a nuisance, and conspiracy with the pulp and paper mill to downplay the dangers of dioxin (Bowman and Toa 1993). As was the case in *Hodges v. Buckeye Cellulose Corporation* (174 So. 2d 565: 1965), the case was thrown out by the District Court because the "plaintiff did not have sufficient standing in court to bring this suit contesting the constitutionality of Chapter 24952."

Local tensions came to a head in April 1992 when a local environmental activist was brutally beaten and raped by three men who told her to stop fighting Procter and Gamble (Bowman and Toa 1993). Once again, water conflict over the Fenholloway caught national media attention. This time, the CBS news program *60 Minutes* aired the brutality and harassment by local residents supporting the plant against those who battled against the main employer in Taylor County. The controversy was further publicized when Florida State University's public television affiliate premiered "Troubled Waters" on January 21, 1999. The documentary portrayed the battle that raged between the mill's supporters in the county and the downstream property owners.

Throughout the conflict over effluent discharges, a main problem has been a lack of technical information that all stakeholders considered reliable. The initial battle was over whether chemicals released into the river contained dioxin, and whether the dioxin could harm the public. According to an EPA Environmental

Working Group's Toxic Release Inventory, direct toxic discharges to the Fenholloway between 1990 and 1994 totaled 1.9 million pounds (EPA 2003). It was also around this time that the working group ranked the Fenholloway River as the 30th most polluted water in the United States. In July 1999, EPA alleged that the river had dioxin levels 200 times greater than the agency's standard allowed (Hollingsworth 2001).

Buckeye has historically been an industry leader in treating its effluent and was the first pulp mill in Florida, and among the first in the nation, to establish an elaborate wastewater treatment system, winning an award for initiating the latest technology to mitigate water pollution (Bowman and Toa 1993). In the 1990s, Buckeye denied that the Fenholloway was contaminated with dioxin. Michele Curtis of Buckeye argued publicly that the plant had stopped using elemental chlorine in its bleaching processes, instead using a less damaging chlorine-dioxide, which breaks down more quickly after use.

Jim McNeal, the groundwater section chief for the Florida Department of Environmental Protection (FDEP), reportedly had said that they "did find contaminants [in tested well water] similar in chemistry to the effluent that was being discharged... Compounds are there, but primary standards for well water are not being broken" (Pfankuch 2002). A report released in 1995 by the FDEP Division of Water Facilities also suggested that "groundwater within the county is quite susceptible to contamination by activities on or near land surface" (FDEP Division of Water 1995). The Report did not mention dioxin, however.

Because of increasing national concern with clean water, the pressure was growing for Florida to reclassify the Fenholloway to Class III, which would require water standards suitable for recreation and habitat for fish and wildlife. Ultimately, Buckeye agreed to meet FDEP's request for an extensive Use Attainability Analysis (UAA) on factors affecting attainment of Class III uses. The technical team came from Buckeye and six engineering and environmental consulting companies (Andreu and Weeden 1996). As a result of the UAA recommendations, and after a string of conflicts relating to the river usage, the Fenholloway River was designated Class III surface water by the Environmental Regulation Commission on December 15, 1994.

According to the UAA, the most feasible method of complying with the more stringent criteria for Class III for Buckeye Florida L.P. required three components:

1. build a pipeline to relocate the discharge point downstream to the estuary (about 1.5 miles upstream from the river's mouth).

2. modify facility processes to reduce the amount of pollutants generated.

3. add dissolved oxygen directly into the wastewater system.

FDEP and Buckeye agreed to implement these recommendations and included them in the NPDES permit drafted in 1997. However, Buckeye's hope to obtain a discharge permit quickly by accepting the Class III requirements was disappointed as the permit application came under opposition from environmentalists, particularly Help Our Polluted Environment (HOPE), American Canoe Association (ACA), and Clean Water Network (CWN)/Natural Resources Defense Council (NRDC). These groups had no role in the development of the

UAA, and did not trust FDEP to represent their interests in the permit negotiations. Despite the opposition, the FDEP issued a draft NPDES permit based on UAA recommendations in August 1997.

EPA oversees NPDES permits, and sent its own team of EPA engineers and contractors to evaluate the plant and the draft permit in January 1998. The EPA Pulp and Paper Technical Team conducted the study (the Tech Report) at the request of the Regional Administrator for EPA Region IV. Based on the technical team's visit to Buckeye, EPA sided with the environmentalists and formally rejected the permit application in March 1998. EPA concluded in its letter to FDEP that "much new technical information has come to light that was not available at the time of earlier drafts of this permit," and listed six objections to the state of Florida's proposed NPDES permit for Buckeye Florida L.P. Among them: the proposed permit contained no color limits, the permit Biological Oxygen Demand might not reflect actual performance capability of the mill, and there was no "Reopener Clause" provision with regard to any future national effluent guidelines for the dissolving kraft subcategory imposed under the Clean Water Act (EPA 1998).

EPA, FDEP, and Buckeye negotiated at length between 1998 and 1999, but the process became irrevocably stalled because of disputes over technical issues. EPA refused to approve the draft NPDES permit containing the pipeline, which moved the effluent without reducing it. The EPA team estimated that, by implementing the oxygen delignification with other recommendations, Buckeye could eliminate the need for a pipeline at a cost of $48 million. But when Buckeye asked BE&K Engineering to estimate the cost of implementing EPA's recommendations, the cost jumped to $97 million (Horkan 1999). Buckeye refused to install in-mill technologies that it argued would be too expensive, would hamper its ability to produce its specialized product, and would not fully restore the river in any case. FDEP was caught in the middle as the agency that prepared the permit based on the UAA recommendations but could not obtain EPA's approval.

Permits and Learning: Limitations and the Hurtful Stalemate

The situation on the Fenholloway in 1999 might be described as a "hurtful stalemate" (see Chapter 19), as administrative and legal actions became hopelessly entwined. The five-year renewal cycle for Buckeye's NPDES would seem to provide a useful device for balancing the mill's interest in stability to facilitate a systematic planning horizon with a limited role for the intervention of interested stakeholders. But the dramatic change created by the river reclassification, combined with the conflict between FDEP and EPA, had led to what seemed an intractable dispute.

Before 1995, Buckeye Florida had to submit *two* permit applications to discharge effluent into the Fenholloway River: for a state Environmental Resource Permit and for a federal NPDES permit. After 1995, both the state and federal permits were consolidated under a Memorandum of Agreement between FDEP and EPA (*Fla. Stat* 403.0885), which established policies, responsibilities, and procedures under federal regulations, defining the manner in which the NPDES permit was administered by FDEP (40 *C.F.R.* Part 123). Buckeye submitted its permit renewal application to FDEP on May 25, 1995; in August 1995 it submitted the Environmental Resource Permit application necessary to implement the UAA recommendations. The second application to renew its NPDES permit was done in 1997 in a "timely and sufficient" manner. As has been discussed above, in March 1998 EPA rejected the FDEP-drafted NPDES permit for Buckeye. Subsequently FDEP had to issue an Administrative Extended Permit to Buckeye, which allowed it to continue discharging industrial effluent in the Fenholloway River based on previous permits until the new permit is approved.

The legal situation was complicated by the refusal of EPA to hold the required public hearings to clarify its objections to the permit. According to the administrative procedure, the FDEP and Buckeye could have requested that EPA hold a public hearing on its March 28, 1998 objections, which would enable the FDEP to modify the permit to address EPA's concerns. Failing that, EPA was required to issue a federal permit to meet the Clean Water Act requirements. Instead, the dispute lay in stalemate.

The environmentalists claimed the Administrative Extended Permit was illegal because it did not meet the statutory Class III designation—the old NPDES permit met only the old Class V water use designation. But they could not file suit against FDEP's decision until EPA held a public hearing.

From the FDEP perspective, EPA's permit rejection was not only an administrative embarrassment but also a failure of confidence. FDEP, too, could not move forward in amending the permit unless EPA held a public hearing. Technically, FDEP had made the appropriate decision within the legal framework of the statutes, interpreting them to allow Buckeye to meet the water quality standards in the Class III water body classification. The FDEP decision was based on the UAA recommendations and credible agreements it had secured with Buckeye to invest in viable technologies stated in the Report. The EPA rejection reflected its lack of confidence in FDEP, but it also subjected FDEP to accusations from the environmental groups that it had sided with Buckeye. EPA's inaction only avoided litigation from these groups against the FDEP in the short run.

Buckeye, for its part, was very dissatisfied. It had cooperated with FDEP to not challenge Class III status and to pursue NPDES permit renewal, yet it continued to face uncertainties over the standards it would be required to meet. It had complied with the statutes when submitting its NPDES permit in a timely and sufficient manner, and accepted the UAA recommendations. No litigation accusing Buckeye of violating the law had succeeded, and the mill had performed significantly better than all established federal standards for existing and new pulp mills in both the dissolving and bleached paper-grade kraft subcategories (Andreu and Weeden 1996). The EPA decision to reject the FDEP draft of its NPDES permit was seen by Buckeye as an indicator that FDEP's regulatory actions could not protect Buckeye against future litigation from environmental groups. Further, all this regulatory inaction had led Buckeye to reluctantly invest substantial funds in additional studies in the hope of developing agency consensus on the appropriate technologies.

Fenholloway River Evaluation Initiative

The Fenholloway River Evaluation Initiative evolved in this atmosphere of administrative stalemate, encouraged by changes in EPA's policies after George W. Bush was elected President. Anticipating these changes, the Clean Water Network's (CWN) Linda Young phoned FDEP Division of Water Quality assistant director Jerry Brooks in May 2001 to suggest a collaborative strategy to resolve the decade-long dispute. Brooks was sufficiently impressed with the idea that discussions began in May 2001 to define a main objective for such a collaborative effort as well as to design a structure. Eventually they signed a Memorandum of Understanding (MoU) dated December 17 and a Memorandum of Agreement (MoA) dated December 21. These detailed a timeline for the Initiative as well as the financial contributions of the key actors towards the Initiative's expenses. The main objective was to evaluate the feasibility of changes in the Foley mill's industrial processes and wastewater treatment system, along with wetlands treatment alternatives. Options would have to be economically and technologically viable and meet Class III water quality standards. The MoU also stated that the Initiative is intended to support the issuance of an enforceable NPDES permit for Buckeye.

A steering committee that included representation from FDEP, EPA, Buckeye, CWN, and NRDC coordinated the efforts of four working groups focused on wastewater technology, process technology, wetlands water quality, and instream water quality. Ground rules for the deliberations were built by consensus prior to the Initiative's formal discussions. These are presented in Table 3-1; they specify representation, decisionmaking rules, the scope of jurisdiction of each working group, and requirements for release of public information from the discussions. FDEP's representative is defined as the "point person" in each working group; EPA's and Buckeye's representatives as experts to provide technical resources. Resource persons not representing the signatory organizations could be appointed, but only with the approval of the signatory organizations. Conflicts were to be resolved at the working group level when possible, but would be sent to the Steering Committee when this was not achievable. All formal decision authority rested with the Steering Committee. The only public distribution of information from the Initiative was to be by joint statement of the Steering Committee.

The Process Technology Working Group began work in 2002 amid controversy over the role of an EPA senior staffer who met with the group but who some argued was not a proper member. This staffer had been instrumental in the 1998 EPA decision that overturned FDEP's permit-granting decision. This controversy carried over into disagreement over the text of the group's final report.

The Wetlands Treatment Working Group spent its first year in data collection and has yet to release a report. The group faced early disagreements over its geographic scope, with CWN arguing for a less expansive area and Buckeye arguing for a more expansive one. Eventually, the Steering Committee found that "if Buckeye continued to wish to pursue a coastal wetlands discharge option, the issue fell outside the purview of the Initiative and would not be pursued within the Steering Committee's mandate" (Seventh Steering Committee Meeting Minutes 2002).

Table 3-1. *Rule Types and Descriptions*

Rule types	*Description*
Position: set of positions and how many participants are to hold each position	• Defines Initiative's memberships and Steering Committee members, i.e., representatives from Buckeye L.P., DEP, EPA, and CWN / Each workgroup has representative from each organization • Defines "point person," "shared objective resource," and "process facilitator" / DEP acts as workgroup's point persons / EPA and Buckeye representatives act as outside expert (shared resource) to provide technical assistance
Boundary: who has access to the decision process	• Members must belong to one of the signatory organizations (MoU and MoA) / Steering Committee can appoint shared resource persons, i.e., Norm Liebergott, Frank McFadden, and Larry Schwartz / Can accommodate other invitees but only through recommendation and agreed by Steering Committee • Members can belong to one or more workgroups
Authority: set of actions assigned to position at each node of decision tree	• Lead representatives are to coordinate with their respective upper management for direction when agreed decision is reached in the Initiative • Internal conflict is resolved at workgroup level and if failed, the Steering Committee will resolve the conflict / If conflict is still not resolved, actors can leave the Initiative
Aggregation: who gets to decide what	• Steering Committee makes final decision only after all facts are presented to the committee • Although no formal rule set on voting, decision is based on general consensus—majority rule • Agreed to have neutral actor provide unbiased opinions to resolve information disagreement, i.e., Norm Liebergott
Scope: scope for each workgroup to work on outcomes that are allowed, mandated, or forbidden	• Assigned workgroup with specific objectives—Process Technology Workgroup / Wastewater Technology Workgroup / Wetland Workgroup/ Instream Water Quality/Modeling Workgroup • To determine availability of economic options to all actors in the Initiative to meet Class III water usage for Fenholloway River
Information: how information is received or disseminated in a timely manner	• All public communication will be released jointly by the Steering Committee / Specify types of information acceptable to all members and how information is to be collected, e.g., for water quality modeling / Information confidentiality is crucial • Correspondence be clearly identified in e-mail subject heading and sent to "Fenholloway Team Distribution List"
Pay-off: how benefits and costs are distributed	• Specify commitment of all parties in terms of time and financial resources / Agreed budget for "shared resources," i.e., service payment for McFadden, Schwartz, Liebergott

It is too soon to report definitive outcomes from the Initiative, but the level of contention in the process does not bode well for the final agreement. On the other hand, each major party has had opportunities to voice objections to prior proposals and to lay out its preferred options. Agreement was reached on the structure and finances of the Initiative, which has built a record of successful joint problem solving among a group of adversaries.

The Challenge of Representation

The normal processing of the state Environmental Resource Permit and the federal NPDES permit, while formally open to public comment, had been effectively hidden from the view of both environmental watchdog groups and Taylor County residents. Yet the two regulatory agencies had made strong efforts to respect both pro-economic and pro-environmental concerns.

The Initiative has changed these dynamics fundamentally. Inclusion of environmental representatives acceptable to CWN and NRDC has greatly increased the time and effort required. But if consensus can be achieved on the permit conditions, court challenge and public opposition is unlikely. By allowing environmental group participation and encouraging transparency, FDEP can no longer be easily painted as taking the industry's side.

Buckeye will be better off if consensus-based permits can be secured. One of Buckeye's objections to the EPA Technical Team's Report was that it failed to understand the existing process technology in the plant. But this was in part due to Buckeye's own secrecy, which sprang from both the need to protect its business interests and its suspicion of the EPA Technical Team's intentions. Opening up its process technology not only to EPA but also to an outside expert and advisor to the Steering Committee has improved the company's credibility in this Initiative.

As for the CWN and NRDC, had they not been able to tap into the technical expertise of FDEP and EPA in addition to their own consultant, they would not have been able to gain access and help shape the decisions about what should be studied in the Initiative's technical reports. In this circumstance, as inside groups, the CWN and NRDC are in a much better position to bargain for their objectives earlier in the process when there is more flexibility.

The degree to which the citizens of Taylor County are now represented in these proceedings is unclear. CWN has North Florida roots, but no Taylor County residents are known to be active in its leadership and its internal procedures for determining positions are not known. On the pro-industry side, whether Buckeye is a suitable representative of wider sentiment in Taylor County is also unclear.

The Challenge of Process Design

Beginning in 2001, great attention was paid to the design of the deliberative process surrounding Buckeye's permit applications. Laborious discussions went into the MoU and MoA, but didn't permanently resolve disagreements over process. Two working groups debated sitting certain representatives, others whether observers were to be permitted at meetings. Many of these debates may be attributed to the low level of trust among the parties before and during the Initiative.

The Initiative has harnessed three factors to overcome intergovernmental fragmentation and the business–environmental ideological divide that had previously led to stalemate. First, it has attempted to resolve disputes over what actually constitutes "acceptable" scientific information. Because all actors have a representative in each workgroup, differences can be resolved before reaching the Steering Committee. Furthermore, the interdependency amongst workgroups allows an internal "check and balance" mechanism to work, especially when presenting their findings at the Steering Committee. This has enhanced the accountability of representatives to the group and encouraged them to reach a solution acceptable to all stakeholders.

Second, credibility is improved when confidential information regarding process technology is shared among the competing stakeholders. When Buckeye allowed the CWN representative to study its process technology, this broke the mill's silence on its current technology's capacity to reduce discharge; the earlier EPA Technical Team did not have this luxury. The current design process has allowed a more transparent and open estimate of factors constraining Buckeye's operation and technology processes. For example, when the Process Technology workgroup failed to report how the technological feasibility, estimated costs, and other factors such as cold caustic extraction were arrived at in the report, the Steering Committee directed the workgroup to redraft its report and evaluate the rationale behind its recommendations. Presence of a few outsiders in the discussions was crucial to full examination and legitimization of the workgroup's decisions, particularly to environmental groups.

Third, the extensive negotiations to resolve the hurtful stalemate require coordination, commitment, and patience of all parties. Scheduling meetings, conducting additional empirical testing, and drafting preliminary reports all drag out the process of building consensus. Although they are needed to complete assigned tasks, they also cause delay in fact-finding and reaching decisions. The Initiative has already consumed more than three years.

The Challenge of Scientific Learning

The Initiative aimed to reach a consensus on the type of technology required. Throughout the conflict over effluent discharges, the main problem has been a lack of technical information considered reliable by all stakeholders. However, during the Initiative Steering Committee Meeting, the dispute over EPA's technical team report resumed when the Process Technology Workgroup (PTWG) presented its preliminary report on process technology improvement options. This was the first positive contribution of the collaboration, and opened up avenues for building consensus on credible data and the scientific methodology to examine them as the Steering Committee attempted to forge an acceptable solution.

One of the difficulties of scientific learning in this case is that each actor in the Initiative must recognize the possible validity of different views from different specialists. The process of learning itself is an instrument for advancing scientific knowledge, but it has to make explicit the critical assumptions on which the policy choices are made. For example, Donald Anderson, who led the EPA Pulp and Paper Technical team evaluating Buckeye production technology in 1998, criticized the PTWG report because he could not determine how some of the technological feasibility, cost, and other factors were arrived at in the Report. To answer this and similar criticism, PTWG agreed to confer further with Donald Anderson in an effort to complete the Report.

Scientific learning is most likely to advance if the process considers options that are favored by all participants. For example, initially CWN insisted and Buckeye and FDEP mutually agreed not to include a coastal treatment option. However, after two years of meeting, the coastal treatment option suddenly appeared in the last Steering Committee meeting, and altered the course of the Initiative. During the Steering Committee meeting, Ray Andreu (Buckeye) pointed out that, if the San Pedro Bay wetland option does not achieve acceptable water quality standards or if potential costs for that option are significant, Buckeye would prefer to pursue the evaluation of other possible wetland treatment options that would be downstream from the existing discharge (Seventh Steering Committee Meeting 2002). In this case, predetermined policy disagreements limited the research agenda in a way that did not allow scientific learning to inform choice.

The Challenge of Public Learning

Prior to the formation of the Initiative, no decisionmaking mechanism satisfied the needs of competing groups. Promoting trust and accountability between two competing views about the best way to resolve the water quality problem was a major challenge. For example, early information related to the river pollution was mostly reported through the local and national media. FDEP was accused of siding with Buckeye with its lax regulatory efforts. On the other hand, Buckeye was suspicious that it would not receive impartial treatment from the EPA since Linda Young's opposition to Buckeye's NPDES permit was supported by EPA administrator Carol Browner, a protégé of Vice President Al Gore. This damaged any trust remaining between Buckeye and EPA.

Given the previous opposition, suspicion, and lack of trust, there was an attempt to confine and resolve disagreement within the Initiative. Yet this process has excluded the local residents and other environmental groups not directly represented. Indeed, through the Initiative, the policy process has evolved into a small number of elite interests representing only some of the affected resource users. The challenge of public learning still remains, i.e. how best to inform the public about the progress of the Initiative. For example, how will the Steering Committee communicate with the public while avoiding the release of premature information or conclusions? Agreeing that all public communication would be released jointly by the Steering Committee helped build confidence, but it left unanswered the question of whether the final outcomes would be acceptable to other groups outside the Initiative. While the DEP, EPA, and Buckeye were the legitimate representatives of their organizations in this negotiation, environmental groups such as CWN were not. This is a key challenge because, as pointed out by Scholz and Stiftel (Introduction), different environmental groups have "different perceptions about the resilience of nature and the rights of property owners and local communities to exploit local resources." Without legitimacy in the decision process, it is uncertain how all relevant views can be represented.

Despite these difficulties, the Initiative was very effective in promoting public learning beyond what was possible in the normal regulatory permit deliberations. Prior to the Initiative, public learning was dependent on the episodic involvement of journalists. The Initiative involved the regulatory agencies, the Foley mill, and at least some environmental groups, so these parties are more likely to understand the magnitude of the problems, the costs, and the recommendations. This common understanding should make it easier for the regulatory agencies to determine appropriate permit terms.

The Challenge of Problem Responsiveness

Although the MoU and MoA do not state that the outcome of the Initiative must be equitable, the MoU does mention that the objective should be reached "in an economically and environmentally sustainable manner, supporting the issuance of an enforceable NPDES permit for Buckeye" (FDEP 2001). We can infer the major differences between CWN/NRDC and Buckeye in the interpretation of "equity" from preliminary interviews we have conducted and from the minutes of the Steering Committee meeting. For the environmental groups equity is a "social justice" issue, in which the Fenholloway River does not belong to the pulp and paper mill nor to the EPA or FDEP. The river belongs to everyone. According to this perspective, those least responsible for creating the pollution are most at risk from its ravages. Linda Young's strong objection to any coastal wetland treatment plan during the Steering Committee meeting is consistent with this view that the plan would only move the discharge location, not resolve the long-term environmental impacts. In her view, the best solution is for Buckeye to invest additional resources for its in-mill process improvements.

By contrast, the "Buckeye view" would see equity in terms of how much it had already invested in its existing technology facilities and the scientific studies its has commissioned to comply with water standards set by the statutes. Buckeye would argue that ignoring available options is "unfair" because leaving any viable options unconsidered only defeats the idea of the Initiative, and thus increases the likelihood of additional investment in in-mill process technology. Furthermore, Buckeye believes that it has not violated any laws by using the river as a point of effluent discharge, and that the objections by EPA and Fish and Wildlife (EPA 1998), the EPA Technical Team Report Recommendations, and the claims that effluent discharge had contaminated well water (FDEP 1995), were neither conclusive nor justified by scientific evidence. Buckeye would not therefore like the costs of mitigation to be added to what it has already done. The "Buckeye view" of equity justifies its interest in managing its production processes with minimum costs of compliance.

Given these differing views, how are we to judge the equity and efficiency of the results of this case? Is there a greater principle for assessing the justice of FDEP's 1998 permit decision, one that can be applied whatever the result? We are tempted to say that if the diverse groups participating in the Initiative can reach a consensus, then an equitable result must be assumed. But this is an argument of convenience, and it presumes that the Initiative will reach an agreement, an outcome that is very much in question at this time.

Conclusion

The Initiative illustrates the challenges of adaptive governance for anticipating, avoiding, and responding to water conflicts. First, instead of fighting over individual preferred options, the Initiative created a process for investigating options based on mutually agreed procedures and criteria. Although there were disagreements over the list of options, the fact that all parties were given the opportunity to express their concerns in the design process prevented the issues from being unresolved outside the Initiative, thus reaffirming each actor's confidence in the decisionmaking structure and reassuring them that no discussion has been limited or stifled.

All actors consider the design process as important as the end product of the Initiative. Inclusiveness at the negotiation table ensures trust and confidence. The challenge of representation in technical proceedings is thus important to all parties. Furthermore, an external consultant trusted by Buckeye's opponents has facilitated a greater level of trust between Buckeye and CWN/NRDC. Although FDEP has been the core actor in coordinating arrangements for the Initiative, the outside consultant has been instrumental in writing technical reports, educating Steering Committee members about in-mill process techniques, and refereeing disputes when an independent voice was needed.

The cost of forming and implementing the Fenholloway Initiative has been high. It has enabled the decisionmaking process to overcome stalemate, but there is no guarantee that the final outcome will achieve stakeholder consensus. The nature of the conflict and the number of actors involved have been challenges. Yet all of the major actors have agreed to participate and contribute time and resources to the Initiative, and it has already advanced representation, scientific learning, and public learning. Crucial to these results, they accept the fact that no option is available unless some form of compromise can be reached. Conclusions on process design and problem responsiveness are yet to be determined.

CHAPTER 4

Tampa Bay Water Wars

From Conflict To Collaboration?

Ayşın Dedekorkut

The "water wars" in the Tampa Bay region between the governments of Pasco, Pinellas, and Hillsborough Counties and the cities of St. Petersburg, Tampa, and New Port Richey began in the early 1970s when densely populated but water-poor Pinellas County started buying land and developing drinking-water wellfields in Pasco and Hillsborough Counties (Figure 4-1). Groundwater pumping in the Tampa Bay region increased 400 percent between 1960 and 1996, and currently over 20 billion gallons of water is exported from Pasco County to Pinellas County every year (Glennon 2002). Pasco and Hillsborough Counties were not happy with the southbound flow of the water. The impact of the wells on lakes and wetlands, including damage to local residents' homes, prompted countless legal challenges, but courts and the Legislature supported the claims of Pinellas County and the city of St. Petersburg. This chapter analyzes two attempts to create institutions to resolve the conflict, one that failed and the current one that holds some promise. We present the history in some detail to demonstrate the complexity of resolving conflicts, and then consider lessons for adaptive governance.

A Brief History of the Tampa Bay Conflict

The First Regional Authority

In October 1974, West Coast Regional Water Supply Authority (West Coast) was created by the Florida Legislature "in response to concerns over negative environmental impacts associated with uncoordinated development of, and competition for, the Tampa Bay region's fresh water sources" (Meinhart 1989). West Coast existed by contract among voting members including Pinellas, Hillsbor-

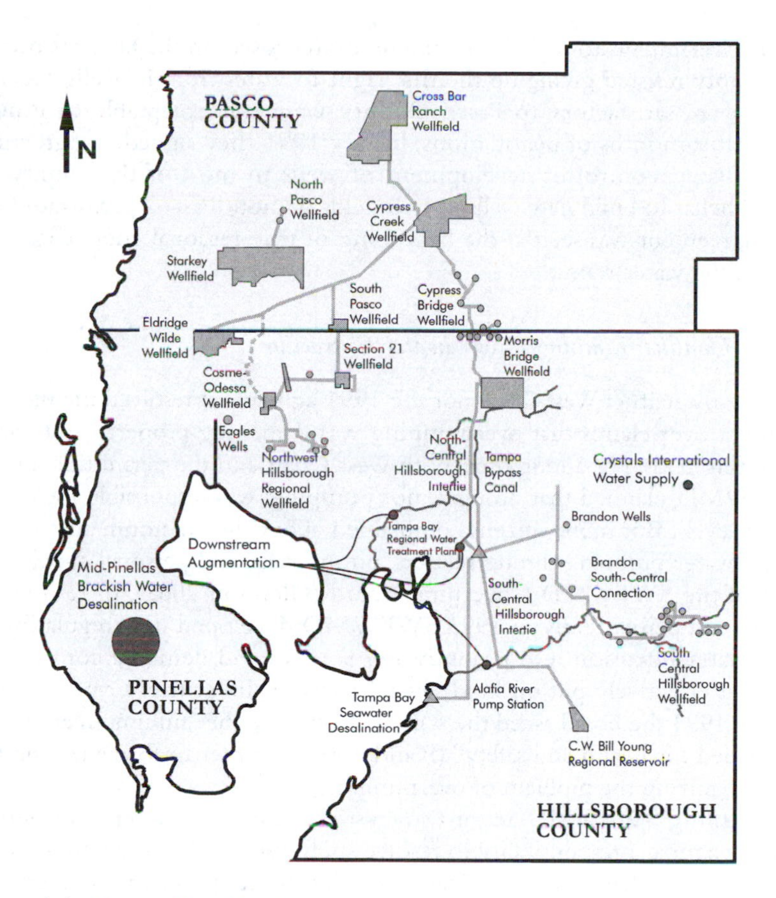

Figure 4-1. *Tampa Bay Region*

Source: http://www.tampabaywater.org/WEB/Htm/About-Us/about-us.htm (Courtesy of Tampa Bay Water).

ough, and Pasco countries, and the cities of St. Petersburg and Tampa. The city of New Port Richey had a seat but no vote. West Coast's board was made up of elected officials or designated representatives from each local government; they agreed to buy water from West Coast, which in turn bought and developed the well fields. The Southwest Florida Water Management District (SWFWMD) and the Florida Department of Environmental Protection (FDEP) were involved as regulating and permitting agencies (see Chapter 1).

Initial Success: Developing a Regional Water Supply Plan

Plans for joint development of future water supplies for the region were constantly hampered by parochial attitudes of West Coast members. Pinellas continued buying well fields in Pasco and held its participation in the $125 million Cypress Bridge well field hostage to pressure the other parties in other issues. Neither that project nor West Coast could continue without Pinellas. During discussions of common ownership of all facilities and a single water rate for all member governments to replace the different rates based on the facilities owned, Pasco County resisted giving up the first right to water from its wells. Guarantees that were satisfactory to Pasco County were not acceptable to Pinellas County. After months of negotiations, in July 1991 they signed an agreement that gave Pasco control of development of wells in most of the county, but allowed Pinellas to build new wells in two fields without Pasco's permission. This historic agreement was seen as the beginning of true regional cooperation and the end of the water wars.

Recurrent Conflict: Limiting Groundwater Extraction

Unfortunately, neither West Coast nor the 1991 agreement resolved the underlying conflict over claims that overpumping was damaging property and habitat close to well fields. For a long time, both West Coast and the permitting authority, SWFWMD, claimed that drought, not pumping, was responsible for declining water levels. But more citizens complained about the environmental impacts of groundwater pumping through letters, phone calls, and personal appearances in front of the SWFWMD Governing Board (Glennon 2002). In response to escalating complaints, early in 1993 SWFWMD developed new regulations to curb saltwater intrusion into groundwater sources and denied a construction permit that effectively put on hold several parts of the critical Cypress Bridge project. In 1994 the board asked the staff to determine the "minimum amount of water needed for health and safety" (Rand 2000) in order to formalize the new policy recognizing the problem of overpumping.

The ensuing SWFWMD actions suggest that Florida's water management districts recognize a responsibility to resolve such disputes, but their ability to do so remains under challenge. West Coast and SWFWMD pushed for conservation as an alternative to new supply development, but area governments continued their longstanding resistance to restricting access to groundwater. Pinellas County leaders in particular worried that water restrictions would kill growth and the economy (Garcia 1993; Garcia and Rogers 1993). SWFWMD then proposed "Water Shortage Orders" to force local governments to take specific actions to reduce water demand (Rand 2000). When West Coast, Pinellas County, and St. Petersburg immediately filed suit against two proposed orders, SWFWMD changed tactics and issued Emergency Order 94-12 in June 1994 to West Coast and its members to reduce groundwater withdrawals, claiming that pumping from well fields had drained lakes and wetlands, killed wildlife, and ruined homeowners' wells. Unlike the Water Shortage Order, an Emergency Order takes effect immediately until a court overturns it. Thus the consequences could be felt immediately to the extent that the concurrency requirements of the 1985 State Growth Management Act prohibit the approval of any new development until adequate public facilities, including potable water, are in place. In September 1994, West Coast sued SWFWMD to rescind the Emergency Order.

These actions triggered the most intensive battles of the water wars: West Coast, its member governments, and SWFWMD spent more than $10 million on legal disputes between 1994 and 1998.

> By March of 1994, every local government and even some of the activists retained counsel and prepared for war. There were in-house lawyers, outside counsel, general counsel and experts on all sides—all paid for with public dollars...[Residents] were paying for at least six lawyers on *all sides* of the case. (Rand 2000) [Emphasis added]

The immediate threat to development and the ensuing legal battles stimulated political intervention. In October 1994 Governor Lawton Chiles formed the Tampa Bay Water Coordinating Council, bringing together SWFWMD and local government leaders to seek a consensus on how to develop and equitably distribute Tampa Bay's water supply to meet existing and future demand while protecting water and associated environmental resources. The Council issued a draft report in early 1995 which recognized the need for new partnerships to resolve differences among the many affected jurisdictions and interests (Jones 1996a).

A truce brokered by Senator Jack Latvala (representative of north Pinellas and west Pasco Counties) led to agreement in July 1995 on a 35-year regional water supply plan as well as a withdrawal of the Emergency Order and subsequent litigation. The plan included more surface water, recycled water, and conservation to reduce use of groundwater. The direct participation of Tampa Mayor Dick Greco and St. Petersburg Mayor David Fischer, along with the absence of Pinellas County Commissioner Charles Rainey (for health reasons), were critical in producing a 4-0 vote in West Coast strongly backing the new plan.

The Coalition of Lake Associations (COLA), representing more than 3,000 property owners in Pasco, Hernando, and Hillsborough Counties, prepared a class-action lawsuit in December 1995 to reduce overpumping and recoup monetary damages from West Coast, Pinellas County, and St. Petersburg as well field operators. The following month, Pinellas County filed a preemptive lawsuit against COLA, SWFWMD, and eight citizens. Pinellas' lawsuit was interpreted as a "Strategic Lawsuit Against Public Participation" (SLAPP) by those who claimed Pinellas sought to use judicial process to intimidate citizens, stifle legitimate public debate, and silence the opposition (Duckworth 1996). To counter SWFWMD's claim of long-term water shortages (backed by a $2.7 million dollar public relations department), Pinellas increased funding for its public awareness campaign from the originally budgeted $300,000 to $800,000 to convince citizens that the environmental impacts on the lakes and wetlands were due to drought, not overpumping (Seaton and Thalji 1996).

The debate of liability was reduced to the interpretation of the word "overpumped," with West Coast, Pinellas, and St. Petersburg defining the term as exceeding permitted quantities, while SWFWMD—which admitted having "overpermitted" (Olinger 1994)—interpreting "overpumping" to mean "too much water being taken from the ground" (Rand 2000). The determination one way or another was critical because of its very different policy implications as well as establishment of liability to pay for the damage and the new supplies.

From a strictly legal standpoint Pinellas' suit was not SLAPP, but some claimed it had the same effect of discouraging public participation (Rand 2000). A published apology in 1999 and the $341,600 settlement for the activists that hadn't settled before might be considered a small victory for the citizens. However, controversy over how to deal with the issue caused the resignation of some COLA board members, and the pressure of the lawsuit resulted in the eventual disbanding of the coalition. These results illustrate the difficulty that affected groups can face when they try to participate in the judicial arena.

In January 1996 the Pinellas County Commission, West Coast, and SWFWMD attempted a joint scientific expert review with the assistance of the Florida Conflict Resolution Consortium to "resolve the scientific issues in dispute involved in the planning and regulatory activities used by SWFWMD to meet the District's responsibility in allocating water withdrawals" (Jones 1996b). A joint panel of scientific experts responded to a list of questions with a report, but it produced no direct actions to resolve how much water each permit-holder could pump.

An administrative hearing prompted by SWFWMD's Emergency Order for four well field permits was scheduled for late July 1996. St. Petersburg, Pinellas County, and West Coast challenged the permits that required pumping to stop if surface water or the aquifer drop below certain levels. One week before the hearing, SWFWMD dropped environment-related conditions from the permit renewal applications, but then denied permit applications because the applicants failed to prove pumping would not harm the environment. This shifted the burden of proof about acceptable levels of pumping from SWFWMD to the applicants. The conflict became so intense that Pasco and Hillsborough appeared ready to pull out of West Coast (Pilla 1996). The hearing issued a Recommended Order directing SWFWMD to renew the permits for the well fields, but a final decision was repeatedly delayed to allow SWFWMD and West Coast to negotiate a deal that would end the controversy.

Legislative Intervention

During the ongoing dispute, Pasco County leaders and others had appealed to the legislature for help in reforming state water policy. In response, the Florida Senate and House of Representatives set up select committees to hold hearings under Senator Latvala and Representative R.Z. Safley. The resultant Water Resources Act was signed in May 1996 after many controversial proposals, including "local sources first" pushed by Pasco activists, were discarded. Most critically, the bill imposed a deadline of October 1997 for SWFWMD to set minimum water levels for lakes, wetlands, and aquifers in priority areas in Pasco, Hillsborough, and Pinellas, and to undertake studies to determine the relative impact of pumping versus drought on water levels. Furthermore, West Coast was directed to consider a series of reforms and report its findings to the legislature the following February.

The deadlines provided critical incentives to resolve the disputes. Worried that if they did not do something on their own the legislature would impose a solution, West Coast officials approved a restructuring in December 1996 that would turn West Coast into a true utility. All members would pay the same wholesale rate for water and share the costs of developing new water sources. This would enable them to develop alternative sources such as a desalination plant. Member governments would turn over their well fields to West Coast and the board would be expanded from five members to nine, three from each county. Making the city of New Port Richey a voting member would increase Pasco County's representation from 20 to 33 percent. All would be elected officials, and a majority would rule. By January 1997, all parties approved the regional Water Supply Plan that included the reorganization of West Coast.

The legislature's October 1 deadline to SWFWMD to set minimum levels and flows put pressure on the negotiations. The toughest issue was Tampa's reluctance to give up its independence from West Coast. Among various special exemptions, Tampa requested unlimited ability to develop new water resources on its own from the Hillsborough River and to continue to own and operate the Tampa Bypass Canal. Pinellas, on the other hand, wanted to control the quality of its water, particularly a veto on the reuse of recovered wastewater. The negotiations dragged on until October, but finally resolved these issues.

Monetary incentives were also critical for achieving a settlement. In March 1997, SWFWMD offered West Coast $325 million over 10 years from the District's New Water Sources Initiative funding to help develop new water supplies. In exchange, West Coast would reduce groundwater pumping to the specified levels and not challenge SWFWMD's environmental protection decisions. This enabled SWFWMD to set minimum flow and level (MFL) rules on September 9, fulfilling its obligation to the state legislature.

After two years of intensive negotiation, all the pieces were in place to resolve the overpumping issue. West Coast's board approved the restructuring plan in March 1998. SWFWMD approved the Northern Tampa Bay New Water Supply and Ground Water Withdrawal Reduction Agreement (the Partnership Agreement) in May 1998. Each member government waived its right to individually develop water supplies and transferred its facilities to the newly formed Tampa Bay Water on October 1, 1998 (Tampa Bay Water n.d.).

Tampa Bay Water

Since its inception, Tampa Bay Water has demonstrated an ability to resolve conflicts in a manner that at least allows progress in building and maintaining infrastructure necessary to mitigate environmental damages while fulfilling water supply obligations to member governments. Conflicts inevitably arise over the location, timing, and nature of individual projects, and meeting the pumping reduction requirements of the Partnership Agreement presents significant challenges to the Authority (Glennon 2002).

For example, the development of a saltwater desalination plant in southern Hillsborough County is vital to Tampa Bay Water's plan to meet the required reductions for pumping groundwater. However, the Tampa Bay National Estuary Program and citizens expressed concern that reducing the amount of fresh water that reached Tampa Bay with the new reservoir project, combined with the discharge of brine from the desalination plant, would dangerously increase salinity in the bay. In November 1999 Save Our Bays and Canals (SOBAC) formed to oppose the desalination plant. They challenged the proposed FDEP permit in order to make the plant dispose of the effluent in another way, such as piping it into deeper waters and installing equipment to raise dissolved oxygen levels in the discharge (Brookes 2001). Mediation sessions failed to resolve the differences and the case went to hearing at FDEP (Swichtenberg 2001). The administrative law judge rejected SOBAC's challenge in October 2001 (Tampa Bay Water 2001) and the desalination plant started operation on March 16, 2003 (*Tampa Bay Soundings* 2003). However, the impacts on the environment are still uncertain since the plant has been mostly inactive since February 2004 due to efficiency problems and will not operate until late 2006 when the repairs will be completed (Membrane and Separation Technology News 2005).

Lessons for Adaptive Governance

Representation

Although not directly involved in the Tampa Bay disputes, elected state officials oversaw regional problems. The legislature and governor both provided alternative institutions for resolving the Tampa Bay disputes. In response to escalating litigation among local governments and regulatory authorities, the Florida State legislature was instrumental in the creation of West Coast Regional Water Supply Authority and later in its transformation into Tampa Bay Water. Governor Lawton Chiles formed the Tampa Bay Water Coordinating Council in October 1994.

Local elected and appointed officials initially reflected the traditional interests of their constituency, so the change from adversarial to collaborative methods in part reflected changing personalities and career incentives. Charles Rainey, Pinellas County Commissioner for over 30 years and West Coast board member from its creation in 1974 to 1996, represented the old guard. "The Pinellas County commissioner virtually founded the authority and acts as if he owns it" (*St. Petersburg Times* 1995). Rainey has been accused of being mostly responsible "for creating such a hostile environment in which compromise and conciliation are not even discussed...confrontational...parochial and litigious approach" (*St. Petersburg Times* 1995) and for using courts instead of other means.

When the winds started to change, Rainey retired two years before the end of his term. "Rainey, who dominated West Coast from its start, staunchly opposed efforts to reduce dependence on well fields or develop alternative water sources, such as desalination. When Rainey left office and West Coast, momentum for a compromise quickly began to build" (Editorial 1998). Rainey was replaced by Steve Seibert, the only Pinellas Commissioner who had opposed the 1996 SLAPP lawsuit.

The restructuring of West Coast and the Partnership Agreement came only after some of the key people who had carved out strong positions left or changed tactics. West Coast General Counsel Ed de la Parte's legal firm had represented West Coast over 20 years, earning over $6 million (Friedman 1994). With the revelation in 1993 that West Coast had the third-highest legal bills in all of state government de la Parte resigned as general counsel, but his status as "West Coast's de facto chief policymaker" did not change because he immediately started representing Pinellas County (Friedman 1994).

Mark Farrell and Pete Hubbell, two top executives of SWFWMD involved in the early conflict, also left. The new director, Sonny Vergara, had ten years of water supply experience, a fact appreciated by the West Coast board. Another new voice was the mayor of St. Petersburg, David Fischer, who replaced one of his city councilmen in the West Coast board and favored the formation of a true regional water utility. First-term State Senator Jack Latvala, whose constituency spanned the two major combatant counties, mediated the dispute between Pinellas County and SWFWMD over the Emergency Order (Moncada 1995b; *St. Petersburg Times* 1995) and played a major role in getting legislative approval for reorganization of West Coast and developing new water supplies.

Design of Decision Processes

Jones (1996a) emphasizes a fundamental structural water supply problem "in that most water issues are regional in nature, while most water suppliers are local." The regional water management districts have regional authority, but the transformation from their original mission of flood control to the water resource protection authority provided in 1973 took considerable time, particularly since existing long-term groundwater permits did not account for environmental quality impacts. Furthermore, SWFWMD's permitting authority provided no means of developing alternative sources of water. Nor did courts provide an effective venue for resolving the conflict:

> Judicial rulings produced mixed results, recognizing the connection between groundwater withdrawals and environmental damage, but not supporting pumping reductions because of the fear of possible consequences resulting from public supply reductions in the absence of any new sources coming on line. (SWFWMD 2001)

West Coast showed promise of finding a regional solution, but its five voting member structure gave Pasco only one vote while Pinellas had two, so Pinellas and St. Petersburg maintained control. Tampa often sided with Pinellas, voting to pump more water from Pasco and Hillsborough; Tampa was generally unaffected, and Pinellas and St. Petersburg's positions made engineering sense to the staff engineer representing Tampa (Rand 2000). Furthermore, local governments could refuse to participate in funding (Rand 2000); Pinellas County regularly refused to pay for alternative source development, which essentially blocked sources other than groundwater because other jurisdictions could not fund these projects themselves.

Compared to West Coast, Tampa Bay Water has the authority, control, and funding to meet its responsibility of providing water in the most cost-effective and environmentally sensitive way possible (Rand 2000). Tampa Bay Water's board provides more balanced representation among regional entities. The nine-member board of directors includes two elected representatives from each member county and one from each city. Of equal importance, Tampa Bay Water has developed a binding arbitration process for counties that disagree with a project. After the arbitration in May 2001 over the new reservoir in Hillsborough County, Tampa Bay Water manager Jerry Maxwell stated that alternative dispute resolution shortened the time to about a third, lowered the cost, and led to a reasonable solution (Heller 2001). The arbitration panel ordered the utility to give assurances that chemically treated water seeping from the reservoir would not damage septic systems or local wetlands.

Despite the structural advantages over the previous authority, Tampa Bay Water has had its share of problems, particularly in terms of institutional memory loss that follows shifts in personnel. In particular, only two holdovers from the founding group took part in the 1998 Water Supply Plan discussion. No new leader emerged to fill the role of former Hillsborough County Commissioner Ed Turanchik, the chair of West Coast during the restructuring, and former Pinellas Commissioner Steve Seibert. The inexperience of new board members in working together, their lack of information about the water problems and alternative solutions, and the reemergence of parochial interests all combined to slow progress in developing the plan.

Scientific Learning

Scientific uncertainty played an important role in Tampa Bay's water wars, particularly in assessing whether groundwater pumping limits were necessary. Throughout the 1980s SWFWMD agreed with West Coast that drought was causing the environmental problems, but a seven-year Water Resources Assessment study started in 1987 demonstrated that the cause of environmental problems in wetlands and lakes of western Hillsborough and central Pasco Counties was not lack of rainfall, but well field pumping. Rand (2000) sees this as a major reversal of policy for SWFWMD. However, convincing the local governments to restrict pumping took considerably longer. Scientific uncertainty remains a source of contention, most recently over the effects of the water desalination plants and of changes in salinity on the bay.

Public Learning

The media was interested in the conflict throughout its long history, and lent support to the claims of Pasco and Hillsborough Counties and citizens about the impact of pumping on property owners and habitats. Rand (2000) notes that during May and June 1994, which she describes as the height of water wars, the *Tampa Tribune* and *St. Petersburg Times* published at least 183 articles on the issue. The *Times* especially pressured both SWFWMD and West Coast to protect the water resources. After changing their position on whether drought or overpumping was causing environmental damage, SWFWMD had the media and the activists on its side and reportedly had 80 percent support (Rand 2000). Pinellas County's decision to invest in public relations in 1995 was an acknowledgement that SWFWMD's campaign was working and Pinellas had to respond in kind.

Throughout Tampa Bay's water wars, facilitative approaches were used several times in ways that extended scientific learning to the broader public. The Tampa Bay Water Coordinating Council in 1994 and the joint scientific expert review in 1996 each produced a report, but the reports were not linked either to public education or to subsequent decisions. However, a facilitator helped the West

Coast board reach a consensus during the restructuring negotiations. Florida Conflict Resolution Consortium Director Robert Jones (2003) claims that although none of these efforts solved the problem by itself, each contributed to the resolution through learning not just about the scientific bases of policy, but also about the alternatives and the costliness of the conflict. Perhaps the heavy costs in litigation and stalled investments provided the strongest incentive for creating Tampa Bay Water: a prime example of the "hurtful stalemates" that sometimes induce adaptive governance.

Problem Responsiveness

Tampa Bay water conflict illustrates the consequences of privatizing a common property resource and the difficulty of negotiating a shift from this system to a jointly managed property rights system. Groundwater has been included in ownership of overlying land under eastern water law, giving the landowner exclusive rights to its use. In the mid-twentieth century, owners' rights in Florida were limited by permit requirements (see Chapter 1). This system of rights was difficult to reform, although changes in permits eventually forced the institutional changes required to develop infrastructure and sustainable use patterns.

Inequity of amount and price of water between different local governments was historically a big problem. Under West Coast each government bought water at a different rate according to which facilities they owned; in some cases Pinellas County paid less for drinking water pumped from the other counties than did those counties. Pinellas County and the city of St. Petersburg systematically benefited from cheap imported water while damaging Pasco and Hillsborough Counties, which had no control over the water exported from their counties.

Different stages of development contributed to the problem. Newly growing Pasco and Hillsborough Counties were water donors while old growth Pinellas was an importer. Pasco and Hillsborough opposed most of the new projects within their borders and proposed a brackish water treatment plant in Pinellas County so Pinellas would contribute its share to the water supply. Pinellas objected to conservation efforts and sources other than groundwater because groundwater was much cheaper, and cheap water was important to sustain growth. Especially throughout 1970s and 1980s, the growth proponents, including developers and builders, dominated Pinellas County. In a *St. Petersburg Times* article Friedman (1994) claims that

> There's no money to be made in conservation ... But there is big money to be made from the construction and technology needed to tap new water sources. And of course, nobody makes more money than the developers who depend on water to lubricate continued growth up and down Tampa Bay.

Others expressed similar sentiments about Pinellas and St. Petersburg's resistance to change. "That's why they have spent millions of tax dollars fighting to keep on pumping. For money. Not for water" (Clarke 1996). "This isn't about water, it's about money. It's about development, and the engine that drives development is water" (Barry 1996).

Initially it seemed cheaper to litigate, but courts never efficiently resolved disputes in the water wars. Major projects to improve the water supply infrastructure were held up. Clearly, the court system imposed very high transaction costs before a negotiated restructuring and regional planning ended the era of unlimited groundwater, the unequal rate structure, and rejection of alternatives to groundwater.

Conclusions

This study of the Tampa Bay water wars highlights several key factors that were instrumental in sustaining, and then ending, the conflict.

Lack of binding authority: Tampa Bay Water Coordinating Council is regarded as a failure due to lack of binding authority, which enabled Pinellas County to withdraw its support when things did not go as they wished. Jones (1996a) argues that many of the recommendations from the Council's report have been incorporated in subsequent plans and agreements, but the Council did not change the local governments' handling of water supply in any fundamental and immediate way. Decisionmaking by unanimous vote resulted in lack of binding authority; West Coast failed to settle the water wars because a single member government could kill a project.

Reluctance to give up existing advantages and independence from other local governments: Pasco's insistence on control of its groundwater, Tampa's reliance on the Hillsborough River for its water supply rather than groundwater supplied by West Coast, and resulting independence from the other governments obstructed negotiations for restructuring West Coast into Tampa Bay Water.

Nature of representation on the West Coast board: The presence of Tampa Mayor Dick Greco and St. Petersburg Mayor David Fischer on the West Coast board, instead of subordinates, contributed to agreement on the 35-year regional water supply plan. Tampa's representation by a non-elected official had created conflict, especially when Tampa voted together with Pinellas and St. Petersburg.

Loss of institutional memory: Key people who had negotiated the restructuring and the Partnership Agreement left the board. With only two holdovers from the initial group in the Tampa Bay Water's board, parochialism and talk of water wars re-emerged during the discussions of the 1998 Water Supply Plan. Pasco's seat on the board changed hands five times within two and a half years of reorganization and interrupted continuity: "The revolving door at Tampa Bay Water is a legitimate concern because newer members, unfamiliar with the aggressive schedule for developing new water sources and reducing groundwater pumping in Pasco, often are accompanied by a 'let's-do-more study' attitude that translates into little more than delays" (*St. Petersburg Times* 2000).

Threat of an imposed solution by the legislature: One factor that drove West Coast's restructuring was members' fear that if they did not do something on their own, the legislature would impose a solution. All agreed that "a solution of their design was better than one imposed from Tallahassee" (Rand 2000).

Financial incentives: The development of alternative supplies was repeatedly thwarted because the local governments could not agree on how to share the cost until the proposal of the Partnership Agreement (*The Economist* 1998).

Change of personalities: The change from adversarial to collaborative approach is attributed to a shift in personalities, especially on the West Coast board, from established officials with careers invested in their positions on water to younger, more conciliatory voices.

Threat of a deadline: The approaching hearing date curtailed the dispute over SWFWMD's 1994 Emergency Order to stop overpumping. Pressure to resolve the issue by October 1, 1997, the deadline set by the legislature to determine minimum flows, was again crucial in moving things along during the discussions to restructure West Coast.

CHAPTER 5

The East Central Florida Regional Water Supply Planning Initiative

Creating Collaboration[1]

Ramiro Berardo

IN RAPIDLY-DEVELOPING Florida, more water has meant more growth, and for politicians in growing towns, more votes—at least as long as water rates could be kept low. Specialized water supply utilities traditionally relied on pumping more groundwater to maintain rapid growth, and had little reason to consider more expensive alternative sources or conservation measures to maintain adequate supplies. This unsustainable use of the natural system calls for adaptive responses from stakeholders to avert a major environmental crisis.

The East Central Florida Regional Water Supply Planning Initiative illustrates the challenges of adaptive governance in the initial stages of collaboration. The Initiative was created as a collaborative response to the environmental problems posed by overuse of the groundwater resource in the East Central area of the State. Figure 5-1 shows the area concerned.

In the past, the St. Johns River Water Management District (SJRWMD) has declared this rapidly-developing area within its jurisdiction to be a Priority Water Resource Caution Area (PWRCA), in which the existing and anticipated sources of water and conservation efforts may not be enough to supply water for all existing and future needs and to sustain the water resources and related natural systems (Vergara 1998). The Initiative seeks a consensual decision process to meet the legal planning requirements while minimizing conflicts and environmental degradation like those in Tampa Bay in the last three decades. The chapter begins with a description of the formation and development of the Initiative, and then considers how the Initiative confronts some of the challenges to adaptive governance.[2]

The Initiative: Creating the Conditions for "Water Peace"

The main water source for agricultural, commercial, domestic, and recreational uses in the region is the Floridan Aquifer, one of the most productive aquifers in

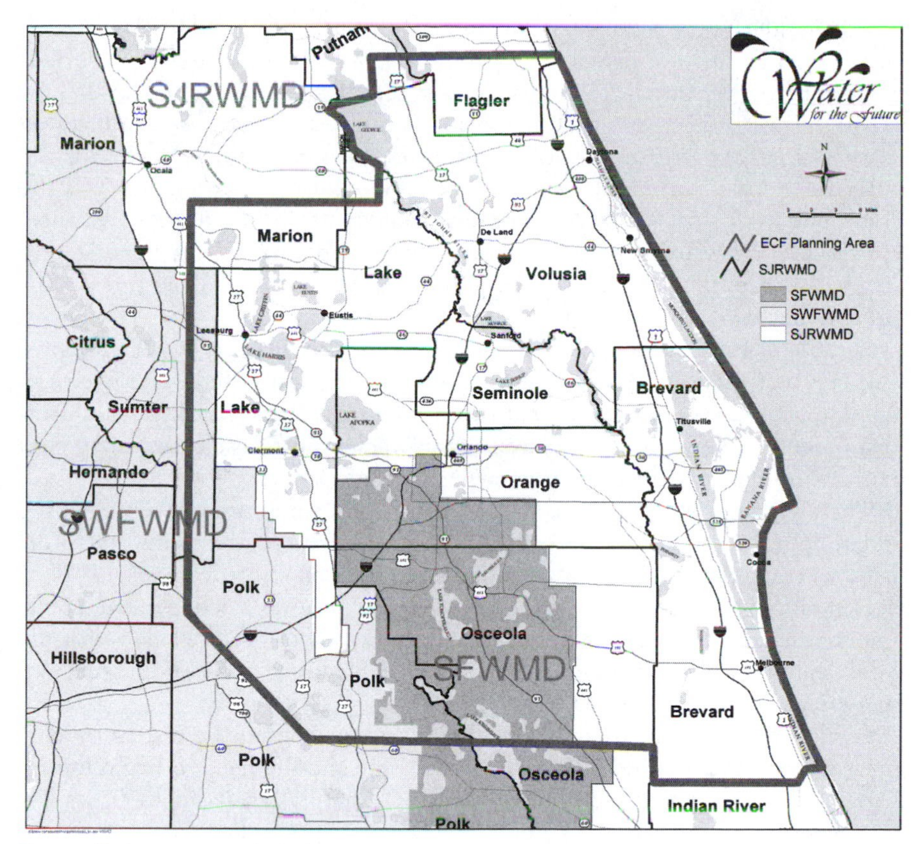

Figure 5-1. *Area Included in the Initiative*

Source: St. Johns River Water Management District (Courtesy of SJRWMD).

the world, which covers over 100,000 square miles in Alabama, South Carolina, Georgia, and Florida (Purdum 2002). Consumption in East-Central Florida is expected to rise from 567 million gallons per day in 1995 to 926 million by 2020, with the share used by public utilities rising from 60 to 73 percent during this period (SJRWMD n.d.). Rapidly growing consumption threatens the ecological equilibrium in the area, creating problems like invasion of salty water into the aquifer, reduction of spring flows, and drying of lakes and wetlands (Vergara 2000).

In Florida the Water Management Districts play a crucial role in the allocation of groundwater. They issue Consumptive Use Permits (CUPs) required for permit holders to withdraw water from surface and ground sources (Chapter 373, Part II, Florida Statutes). As noted by Purdum (2002), "the allocation system is designed to (1) prevent waste, (2) provide certainty to existing users, (3) provide equal rights irrespective of economic power, (4) protect natural resources and (5) provide future users by requiring water managers to address comprehensive planning and resource development." From its oversight of CUPs, the district became concerned over the future water supply in the East Central region, and it was the natural leader in the creation and development of the Initiative.

Stakeholders that formally take part in the process include the Florida Department of Environmental Protection (FDEP) and the three water management districts (WMDs) with authority over the area—the St. Johns River District is the primary actor, but the South and Southwest Districts are also involved to a lesser extent as groundwater boundaries do not conform to the surface water boundaries on which districts are designed. Organizations such as the East-Central Florida Regional Planning Council, the Volusian Water Alliance, and the Brevard Water Supply Board have also played a role in the negotiations. Last, but perhaps most critical, a large group of cities and the public supply utilities that they own are involved.

As early as 1989 SJRWMD began studies to identify Priority Water Resource Caution Areas (PWRCAs), and initial assessments found potential problems in rapid growth areas of Seminole, Volusia, Lake, Orange, Flagler, St. Johns, and Brevard Counties. New assessments extending predictions to 2020 provided extra information about potential environmental damage (Vergara 1998).

These pessimistic forecasts made clear that strong commitment to protect the underground resource would be needed to protect the aquifer from uncoordinated exploitation. SJRWMD organized a workshop in November 1997 for elected officials to initiate the Water 2020 Project. This was a "cooperative public process" created to obtain a more accurate description of the challenges that the East Central region faced and to "ensure that planning was conducted in an open public process" (SJRWMD 2000).

From April through August 1999 the district presented the results to local governments, stressing the importance of the issue and the need to begin immediately the lengthy process to find new sources of supply. Finally, in 2000, the district reviewed calculations of water availability based on recent applications for permits and renewals, with the new forecasting predicting shortages by 2006 instead of 2020. In 2001 the district reaffirmed its protection of the aquifer and stated that extended permits would be issued only to suppliers that implemented measures to protect the groundwater source.

Prompted by the seriousness of the problem, Richard T. Crotty, chairman of Orange County, called and hosted two meetings in Orlando in the beginning of 2002 that were the foundations for the Initiative. The first of these "water summits" took place on January 31, 2002. As stated by Crotty in his opening speech, the objective of these two meetings was to identify cooperative solutions and avoid unnecessary conflict among public supply utilities that could result in expensive and time-consuming litigation (FCRC 2002e). The meeting brought together more than 150 participants who agreed upon a number of points, including:

- equitable distribution of costs, control, and management of new regional supplies,
- a partnership between local governments and public supply utilities, and
- long-term planning by all the operators in the area.

By the second meeting on February 28, 2002, the three water management districts had hired the Florida Conflict Resolution Consortium (FCRC) to help set a formal agenda by November 2002. Robert Jones, Director of the FCRC, introduced the Facilitation/Assessment Team, which included Rafael Montalvo from FCRC and Linda Shelley and Jake Varn from Fowler, White, Gillen, Boggs, Villareal & Banker, P.A. The team conducted interviews in March and April to clarify the interests and motivations of the actors involved in the exploitation of the resource and provided an important foundation for a round of meetings that began in May 2002.

One goal of the Initiative was to avoid a "water war" like the one that had taken place in the *Tampa Bay* area (see Chapter 4). The Initiative was determined to reach agreement without squandering funds in prolonged court battles.

Phase I: Identifying Issues and Setting Goals

On May 22–24, the first of three rounds of workshops divided participants into five subregional workshops. Each workshop identified water supply priorities. "Water Reuse and Conservation were ranked within the top three issue areas for all of the five workshops" (FCRC 2002a). A third important priority area identified was the creation of New Water Supply initiatives.

The second round of workshops on July 10–12 provided participants with a set of statements addressing each of the three main priority areas identified in the first round. Statements that received strong support included:

- Recognize the importance of reclaimed/reused water to: offset the use of potable, high-quality ground and surface water, recharge the aquifer, and mitigate effects of withdrawals from the aquifer;
- Highlight "intergovernmental coordination and cooperation to maximize and optimize the use of reclaimed waters" (FCRC 2002b);
- Emphasize the role of WMDs in providing financial incentives for reclaimed water projects that are part of the regional water supply plan;
- Provide information regarding water conservation, etc.

The third round of workshops on August 22, 27, and 30 attempted to "refine promising, acceptable strategies that might be jointly developed at the subregional and regional levels in a possible Phase II of the dialogue process" (FCRC 2002c). For these discussions, the facilitation team added three new goals: linking land and water development, coordinating intergovernmental relationships, and enhancing water recharge in the aquifer. For each of the now six goals, workshop participants discussed a number of options, which were later presented by the facilitation team as general policy recommendations.[3]

Phase II: Developing Projects

During the summer of 2002, the second stage of the Initiative attempted to strengthen the communication process and develop a list of priority projects that could be included in the district's Regional Water Supply Plan. The geographic region of the Initiative was slightly reduced and subregions were reorganized into four more consistent groupings following the recommendation of the facilitation team, which now consisted of Shelley and Varn from the original team.

The first round of meetings particularly emphasized new water supply projects, along with reuse of reclaimed water and conservation of the resource. The second round concentrated on the link between land use and water supply planning. The third round focused on particular needs and development of projects in each of the subregions.

At the end of Phase II the main focus had narrowed to developing new sources of water supply and linking land use planning with water supply planning (SJRWMD 2003). Eleven new water supply projects were identified for incorporation into the District Water Supply Plan (DWSP) update in 2004. Seven of those involved surface water from five different locations along the St. Johns River, the Taylor Creek Reservoir (a tributary of St. Johns River), and the lower Ochlawaha River. The remaining four projects contemplated three seawater demineralization locations (two in Brevard County and one in Volusia) and the augmentation of the reuse capacity of Lake Apopka.

To link land use planning with water supply planning, the legislature in 2002 required that local governments responsible for the water supply facilities within designated PWRCAs must create a Water Supply Facilities Work Plan (WSFWP) by January 1, 2005. These work plans must be included in the local Comprehensive Plan,[4] and must identify water supply needs and "the water supply facilities and sources of water that will be needed to meet those needs" for at least 10 years (SJRWMD 2003). The district held workshop presentations during the second phase of the Initiative to instruct local government officials and water supply utilities on the requirements posed by the new legislation.

SJRWMD, together with Initiative participants, also developed a "Potable Water Availability" worksheet to help local governments "to identify water supply availability considering both infrastructure and permitted allocation under consumptive use permits" (SJRWMD 2003) when they need to amend their comprehensive plans.

At the end of 2003, the district recommended that the Initiative should emphasize the educational work on water conservation and reuse, the promotion of regional and subregional cooperation among service providers, and the prioritization of the identified projects for new water supply (SJRWMD 2003, *38*). The development of the Initiative so far leaves important lessons that are summarized in the remaining part of the chapter.

Challenges to Adaptive Governance: Contributions of the Initiative

The Initiative illustrates four of the challenges to adaptive governance: representation, process design, scientific learning, and problem responsiveness.

Bringing Stakeholders Together: the Challenge of Representation

A primary objective of the Initiative was to secure the participation of elected officials from local governments because "these are the individuals that ultimately have the power to introduce modifications to the system, to embrace or stop a policy."[5] While utility managers and agency experts had historically controlled the provision of water, the novel challenges of environmental protection could not be met without the participation of political leaders, since they ultimately determined how much should be invested and who should pay for alternatives to groundwater. By encouraging direct interaction among these decision-makers, the organizers of the Initiative offered stakeholders a new way to negotiate their positions without the costly disputes that characterized the Tampa Bay water wars.[6]

However, maintaining participation was not easy. While a large number of elected officials attended the first summit in Orlando, turnout dwindled in the following meetings, with elected officials mainly sending their staff to the workshops.

One explanation is that elected officials are not fully convinced about the extent of the crisis and the need for solutions to the overuse of groundwater. One member of the SJRWMD interviewed for this study said that "they just don't think that the problem is that bad, and as politicians they need the problem to explode in order to do something." Another participant observed "the temporal horizon needed to solve an environmental problem is different than the temporal horizon needed to get reelected. If you want to avoid the problems with water use in Florida, you have to think 15 or 20 years ahead, but if you want to get reelected, you just have to think about the next 2 years." This difference of perspectives could threaten the success of the Initiative.

A second explanation offered in the interviews is that some cities in the region just prefer the status quo, since they already have long-term permits. They did not show up in the meetings because they might be forced to change a water system from which they benefit. The issue prompted a quick response by the district, which adopted an aggressive policy in seeking the stakeholders' participation. It has held presentations and workshops for elected officials, city and county commissions, and a number of organizations with interests in water supply throughout the area covered by the Initiative (SJRWMD 2003). Thus actors not attending the formal meetings are still informed of the problems and proposed solutions. The regional authority is attempting to avoid granting only extremely restrictive CUPs that may provoke providers to go to court; that might initiate political battles that could end in stalemate and lack of commitment to find cooperative solutions to the problem.

Hearing All Voices: The Challenge of Deliberative Process Design

From the beginning the Initiative sought to take into account the needs of interested groups. It actively solicited the opinions of local governments. The solution to the problem will require continued long-term planning and technical development of alternative sources of water. This commitment is more likely if the stakeholders believe they have a voice in the evaluation of alternative sources.

In an interview conducted in March 2003 (before Phase II started), Barbara Vergara, Director of the Division of Water Supply Management of SJRWMD, observed that the district did not want to force stakeholders to adopt decisions that they did not fully embrace. Instead, the district's staff involved in the Initiative was going to try to involve participants in a dialogue, including their recommendations in the proposals for alternative projects of water supply.

The development of both phases of the initiative seems to have followed those lines. In Phase I, each of the six areas around which the process developed was identified as important by a large number of participants in the meetings. The same thing can be said about the goals and strategies elaborated for each of the areas: they were determined by the level of agreement among the participants.

Phase II in 2003 proceeded in the same way, letting the participants' opinions determine the tasks to be addressed in the future. At the end of the workshops in 2003, 11 projects were to be incorporated into the District Water Supply Plan update in 2004. Seven were presented by the district to the participants in the first round of meetings in Phase II. The remaining four were direct proposals of participants in the second and third rounds, who sought SJRWMD's assistance in obtaining technical information and support in order to include them in future amendments of the DWSP.

Participant opinions influenced the exclusion as well as the inclusion of projects on the final list; two projects proposed at the beginning of Phase II were removed. One of them contemplated the extraction of surface water from the lower Ocklawaha River in Marion County, but the Marion County Commission asked to exclude the project from the 2004 update and conduct additional evaluations in the projected site before deciding on the proposal. The second project sought to use Lake Harris as a supplemental source for reclaimed water, but the City of Leesburg, the only potential beneficiary of the project, showed no interest.

Summing up, the collaborative approach designed by the creators of the Initiative has kept the process moving in a direction of understanding and cooperation, which increases the chances of finding alternatives that can accommodate both human development and environmental sustainability.

Sharing Information: the Challenge of Scientific Learning

Widely-shared information was important for developing a common understanding about the effects of an indiscriminate use of the aquifer. During both phases of the Initiative, "water supply issue information was provided to elected officials, water supply utilities, the public and the media through direct mail, the district's quarterly magazine..., the district's monthly local government newsletter..., the Florida Chapter of American Planning Association newsletter, media interviews, and the district's web site" (SJRWMD 2003).

Challenging the scientific basis for decisions is a common practice in environmental politics (Adler et al. 2001). Nevertheless, the studies and information provided by SJRWMD have not been seriously challenged by participants in the Initiative. A member of the facilitation team pointed out that "it was a little bit surprising that they didn't say 'we don't buy your science.' That would have made the process much slower."

This absence of challenges may be explained in three ways. First, before the Initiative, local governments had invested few resources in developing a capacity to challenge the agency. Local governments usually limit studies to specific issues and the consequences of actions in their own jurisdiction. The district has a double advantage: the agency funds research activities from its own tax base, including scientific personnel and a budget for projects,[7] and the "science" that it produces focuses on the regional effects of interaction between humans and the natural system. The scope of the district's science makes it difficult for local stakeholders to oppose the scientific findings, even when they do not share the conclusions derived from those studies.

Second, the district has provided assistance for studies on a variety of issues requested by local stakeholders. Those stakeholders who benefit from the financial and technical support of the district in the scientific assessment of their problems are not likely to challenge it when it offers clear incentives to work cooperatively.

Finally, the Initiative seems to have strengthened collaborative bonds between the district and the local governments. The meetings have helped clarify issues like the dates and conditions for presenting the municipalities' Water Supply Facilities Work Plan, and provided the opportunity for them to work together to present sound alternatives to groundwater use.

Efficiency and Equity: the Challenge of Problem Responsiveness

If environmental protection were the main goal of local governments, public suppliers could reduce the amount of water withdrawn from the aquifer and therefore avoid conflict with the district over permit renewals by simply raising rates. In that case, the price system would provide a straightforward mechanism to protect the source. But politicians answer to voters, who happen to prefer low rates and large amounts of water. Elected officials would agree to look for alternative water sources only if this did not harm the financial health of their constituents. This is extremely difficult when water from any other source is more expensive than water from the aquifer. Note, however, that a key problem underlying this structure of incentives is that users are not widely "educated" about the negative impact of resource overuse. Hypothetically, suppliers could instruct users on the negative impacts of excessive withdrawals and then reward (or punish) them with incentives to conserve water. As long as this problem is not solved, the tension between suppliers and the authorities—cheap water from the aquifer versus environmental sustainability—will persist.

Another problem arises because permit applicants perceive that the permit granting process is inequitable, or that they cannot obtain the most favorable permit when "others seem to get what they want."[8] One county utilities official complained that SJRWMD did not take into account that some cities invested more than others: "We have invested millions of dollars to obtain the water, but we are getting a worse permit than places that did not invest even a half of what we did....[T]his system is obviously not fair." SJRWMD acknowledges that these perceptions of unfairness threaten regional cooperation, and they expect the Initiative to change these perceptions by disseminating information and helping local providers share their experiences. To accomplish this, the Initiative needs a high level of stakeholder participation, and must explain just why some providers get better permits than others.

Conclusions

The East Central Florida water supply initiative was an attempt to overcome the collective action problem that arises when multiple users share a finite source of groundwater. I have identified the stakeholders and suggested how this cooperative effort may overcome the challenges to adaptive governance. The initiative provides mechanisms for dealing with the collective action problem: a participative environment, the diffusion of information, and collaborative efforts between local governments and the regional authority (SJRWMD). So far, the collaboration has moved slowly but deliberately. The regional water authority has learned from previous experiences in other areas of the state that the simple application of its permit granting authority does not lead to efficient outcomes when the relationship with stakeholders is not characterized by trust and a disposition to work cooperatively.

However, the process faces a number of challenges that, if not solved, could lead to the failure of the initiative. The difference in incentives for cooperation is the main obstacle. While the goal for the district is to protect the natural system from excessive withdrawal of groundwater without using draconian controls, for the suppliers the main objective seems instead to be economic efficiency and guaranteed supply. Elected officials have to face their constituency, and they want to provide the highest possible benefits without imposing visible costs. This means taking water from the source that is cheaper: the aquifer. If other sources have to be found, local officials want financial incentives that will allow them to find these new sources without jeopardizing the financial health of their communities (FCRC 2002d).

This situation makes us cautious regarding the future success of the Initiative. Cooperation sustained during the relatively painless process of dialogue will face increasing strains as firm commitments are required. Consensual institutions can produce great levels of agreement without necessarily establishing a clear will to cooperate (Lubell 2002b). If the Initiative does not produce results soon, stakeholders may conclude that "cheap talk" leads nowhere, which could hinder efforts to avoid an East Central Florida Water War in which individual water suppliers contest restrictions by the regional authority whenever their financial interests suffer.

The valiant efforts of the district to direct negotiations, and its willingness to accommodate myriad divergent views, may be insufficient to secure cooperative negotiations. While it is true that new sources of water supply are needed, it is also true that local governments must educate users on the protection of the resource. It is only through the complemented action of all actors—governmental and non-governmental—that the future of water quality in the East Central area of Florida will be secured.

Notes

1. I am grateful to John Scholz, Bruce Stiftel, and the students in the Water Conflicts seminar at Florida State University for the rich exchange of ideas that has improved this piece. I also extend my appreciation to three anonymous reviewers for their helpful comments.

2. In addition to the listed references used to research this piece, 14 phone interviews were conducted with key informants in the area, including top officers of SJRWMD, members of the Initiative's Facilitation Team, and local officials involved in the process.

3. http://www.sjrwmd.com/programs/acq_restoration/res_devel/ecfla/fs_ECFwatsupply%20.pdf.

4. The 1985 legislature adopted the Florida's Growth Management Act, which requires all municipalities and counties in the State to create local Government Comprehensive Plans to manage future growth. These plans address issues such as coastal management, conservation, and of course land use (Fla. Stat. ch. 163, part II).

5. Robert Jones, personal communication.

6. Ostrom (1992, *302*) notes that "the capacity to think about alternative coordinated strategies is affected by (among other things)...knowledge about the experiences of other groups trying to solve similar problems." In this sense, the case of Tampa Bay was especially important for the emergence of the Initiative, since the "Water Wars" in that area cost millions of dollars and the environmental damage was tremendous.

7. The district's budget for the fiscal year 2003–2004 allocated $14,401,046 to water resources planning and monitoring. From this amount, $6,276,122 is directed to district water supply planning, while the remaining amount ($8,124,924) serves the purpose of research, data collection, analysis, and monitoring.

8. In the interviews conducted by the FCRC between March and May 2002, one of the statements upon which interviewees showed a high level of agreement was "Our consumptive use permit is not as 'good' as other local governments': the district is making these decisions arbitrarily" (FCRC 2002d, *8*).

CHAPTER 6

Apalachicola-Chattahoochee-Flint Basin

Tri-State Negotiations of a Water Allocation Formula

Steven Leitman[1]

AFTER DECADES OF conflicting demands by competing water users, the states of Alabama, Georgia, and Florida attempted to negotiate an interstate water allocation formula for the Apalachicola-Chattahoochee-Flint (ACF) basin between 1997 and 2003 as part of a tri-state compact. The ACF case illustrates the lessons to be learned from the failure of interstate negotiations and the challenges to adaptive governance.

Water Users and the ACF Basin

The Apalachicola River is formed by the confluence of the Flint and Chattahoochee rivers, as illustrated in Figure 6-1. The Flint and Chattahoochee are very different in nature and in usage. The Chattahoochee's source of flow is primarily surface water, and multiple storage reservoirs allow the basin's water resources to be managed. The Flint River, on the other hand, has a large groundwater flow and almost no reservoir storage capacity. Therefore, flow in the Chattahoochee basin can be managed by regulating both supply and demand, whereas flow in the Flint can be managed only through demand. Management of the Flint is also complicated by the surface–groundwater interactions in the mid-to-lower Flint basin. At median flow the Chattahoochee River typically provides slightly more water to the Apalachicola than does the Flint, but during low flow, the spring-fed Flint typically makes a greater contribution.

The major water diversions from the Chattahoochee River are to municipal supply, while the major diversions from the Flint are to agricultural irrigation. The reservoirs in the Chattahoochee basin are managed for municipal and industrial water supply, electricity generation, waterborne transportation, flood

Figure 6-1. *Apalachicola-Chattahoochee-Flint River Basin*

control, wildlife management, and reservoir-based recreation. Evaporation losses from both main stem reservoirs and other impoundments in the ACF watershed are a major source of water depletion, especially during the summer when they can exceed consumptive depletions by metropolitan Atlanta. Relative to flow in the Apalachicola River the storage capacity of the reservoirs is limited and therefore so is the capacity of these reservoirs to augment downstream flows. Relative to flow in the Chattahoochee River at Atlanta, however, the storage capacity is large, leading to differing perspectives of the capacity of the reservoir system between upstream and downstream interests. Political resistance to lowering water levels in the reservoirs during summer low-flow season, from waterfront homeowners and recreational users, complicates management of the reservoir system.

The conflicts that spawned the ACF Compact were more of a divergence between upstream and downstream interests than differences between the three states per se. From the start there was a disparity between how the negotiations were divided (i.e. by state borders) and how the interests of the stakeholders involved were divided. This discrepancy was most pronounced in Georgia, where about three-fourths of the basin lies. Many interest groups in Georgia downstream of Atlanta or in the Flint basin were more aligned with the interests of downstream states while the negotiating positions of Georgia tended to be dominated by the upstream interests of Atlanta.

Each state sought to dictate the terms under which the U.S. Army Corps of Engineers would release water, rather than seek to apportion the waters of the basin as a shared public resource. Georgia wanted to retain as much water in the federal storage reservoirs as possible to support continued growth in Atlanta, with no restrictions on future diversions other than the physical limits from the source of water in Metropolitan Atlanta. Alabama sought continued water flows for hydroelectric power, municipal supply, and navigation access to inland ports in Alabama. Florida wanted to assure the variable flow of water into Apalachicola River and Bay to support the seafood industry and the estuarine and riverine ecosystems.

Water Rights and Authorities

The Corps of Engineers has certain regulatory jurisdiction over all navigable waters and operates the federal storage reservoirs. Other Federal agencies have jurisdiction over various aspects of the water resources of the ACF basin. For example, the U.S. Fish and Wildlife Service has some jurisdiction over fish and wildlife resources, and the Southeastern Power Administration has a role in the production and sale of power from the federal dams.

Water not contained in the federal reservoirs is subject to riparian rights as modified by each state. All three states use what is commonly referred to as a regulated riparian approach based on the reasonable use doctrine. Nevertheless, one of Georgia's basic tenets in the negotiations was that because most of the basin is located in Georgia, most of the water belongs to it, regardless of the effect on downstream riparian water users.

The three possible venues for resolving water disputes among states are an interstate compact, Congressional apportionment of waters, or a hearing before the Supreme Court of the United States (Ruhl 2003). The U.S. Supreme Court has constitutional authority over controversies between two or more states. A dispute between two states that use the same doctrine regarding their water rights can be resolved through local law. Where the laws of the states differ, or the Court decides that the local law will leave one state inequitably disadvantaged, it can base its judgment on equity rather than local rules.

In *Kansas v. Colorado*, 206 U.S. 46 (1907) the Court held that the two states were entitled to an "equitable division of benefits" from a river. In 1930 the Court expanded the doctrine of equitable apportionment by asserting in *New Jersey v. New York,* (283 U.S. 336, 342 (1930)) that "the effort always is to secure an equitable apportionment without quibbling over formulas." On the other hand, the court limited the doctrine in *Connecticut v. Massachusetts*, 280 U.S. 523 (1929) by requiring that the complaining state adhere to a higher standard of proving injury. The complaining state must show clear and convincing evidence of a substantial injury to its interests as a result of another state's use of the resource, as in *Kansas v. Colorado*. The question that still looms in the background of the ACF negotiations is whether Alabama or Florida could meet this standard in their dispute with Georgia.

With the termination of the ACF River Basin Compact discussed below, the ACF basin may become the first major interstate apportionment case the court has entertained in the era of mature environmental statutory law. It is not clear how thirty years of environmental awareness and regulation may affect the Court's demeanor when it comes to interstate water allocation (Ruhl 2003). The

Court recently indicated a willingness to consider environmental (fish and wildlife) uses in its harm–benefit comparison in *Nebraska v. Wyoming*, 115 S. Ct. 1933 (1995). In seeking to modify an apportionment decree of the North Platte River, Nebraska was permitted to amend its pleadings to add injury to fish and wildlife habitat.

Background to the Negotiations

A contentious relationship among competing water users in the ACF basin extends back to the 1970s due to the limited availability of the federal navigation channel. Upstream interests claimed that limited availability of the federal navigation channel hindered their economic development, whereas Florida refused to accept major structural alterations to the river that would have occurred in Florida.

In 1989 the Corps of Engineers proposed to reallocate water in Lake Lanier from hydropower to water supply for the Atlanta area and to formalize current reservoir operations in the form of a Water Control Plan (USACE 1989). Upstream interests reacted by saying that their reservoirs were being used too much to support downstream needs, and downstream interests reacted by saying that too much water was being retained upstream. As a result of including reservoir operations in the proposal, attention expanded to the entire watershed instead of just the upper portion of the Chattahoochee basin. There was a widespread fear that Atlanta's water use would dry up the river, a fear that persists to this day even though Atlanta lies in the upper portion of the watershed and the volume of water it consumes is a small fraction of the water available at low flow in the lower reaches of the basin.

In response to this proposal, Alabama sued the Corps of Engineers for failing to meet the requirements of the National Environmental Policy Act in their preparation of the Environmental Impact Statement. With Florida poised to enter the suit on the side of Alabama, and Georgia on the side of the Corps of Engineers, the three states and the federal government negotiated an agreement to stay the suit and conduct the ACF Comprehensive Water Resources Study. The Comprehensive Study provided technical information, tools to evaluate water resources from a system-wide perspective and background information on the management of river basins. When the Corps of Engineers proposed the reallocation and Water Control Plan revisions of 1989, it contended that it had adequate knowledge to make a decision to reallocate water. Eventually, it took eight years and over $20 million in state and federal funding to pull together the information necessary to begin the negotiations for allocating the waters and defining reservoir management practices.

The Comprehensive Study led to the Apalachicola-Chattahoochee-Flint River Basin Compact (vember 11, 1997. This Compact established the ACF Basin Commission and required it to "establish and modify an allocation formula for apportioning the surface waters of the ACF basin." The Compact did not include specific details of an Allocation Formula because the Comprehensive

Study showed that the Allocation Formula would need to be changed over time. Had the Allocation Formula been included in the Compact, new legislation would have to be passed through three state legislatures and the U.S. Congress to make changes. The legislative difficulty of changing other compacts has resulted in litigation before the Supreme Court (Kenney 1996). If the Allocation Formula were delegated to the control of the three states, the Formula could be amended without legislative approval.

The Allocation Formula was negotiated through the ACF Commission. The Commission consisted of the three governors, with actual negotiations to be handled by their appointed representatives. If an agreement could be reached, the federal commissioner (appointed by President Clinton) had 245 days to either accept or reject the agreement. This decision was to be made with the help of a team of federal agencies including the U.S. Army Corps of Engineers, Environmental Protection Agency, Geological Survey, Fish and Wildlife Service, Southeastern Power Administration, Natural Resource Conservation Service, National Park Service, Forest Service, Maritime Administration, and the National Oceanic and Atmospheric Administration.

The participants in the negotiations fell generally into the following groups:

- The three states which had control and power over the process;
- The federal commissioner who had qualified veto power over any agreement reached by the three state Commissioners;
- Entities with the technical expertise or financial capacity to influence the negotiations (e.g. federal agencies such as the Corps of Engineers, influential private interests such as the Southern Company, powerful governmental bodies such as the Atlanta Regional Commission);
- Entities closely involved with the negotiations, but with less influence on the process (e.g. Tri-state Conservation Coalition, Lake Lanier Association, City of Columbus, the Nature Conservancy); and
- Other stakeholders relegated to a role of observing the negotiations and reacting to decisions.

This set up a hierarchy of spheres of influence among the stakeholders in the negotiations. One of the reasons the negotiations never resulted in an agreement was the limited level of communication and interaction among these spheres.

The ACF Water Allocation Conflict and Challenges to Adaptive Governance

Adaptive governance can be thought of as a governmental process that encourages adaptive human behavior in response to limitations imposed by the interaction between human and governmental systems. For the balance of this Chapter, I will relate the ACF case to the challenges to adaptive governance.

Representation

It is critical to identify and involve groups vital to the deliberative process, and to determine which are likely to forward the process and which are not. Although there was broad-based participation by stakeholders at various stages of the Water Allocation negotiations, the question of whether this participation was adequate warrants further discussion. On a continuum from token involvement to meaningful participation, the nature of stakeholder participation affects whether stakeholders will accept a final agreement.

The level of dedication and time that stakeholders can donate to the process affects the quality of representation, especially in long and complex negotiations. They must have an opportunity to be meaningfully involved and invest the time and energy to make the most of this opportunity. In complex negotiations, stakeholders must be educated to be meaningful participants. When a person or group first enters the negotiation and when changes occur in stakeholder representatives can prove to be important; changing personnel in the midst of negotiations can disrupt a group's effectiveness, delay progress, or consume time educating new members. The timing of entry and continuity of involvement were both continuing problems in ACF negotiations, both for the negotiators and the stakeholders, as might be expected in any multiyear negotiation.

The primary negotiators for each side changed over the course of the negotiations, breaking the continuity of negotiation and the ability of stakeholders to develop trusted relationships with their representatives. In 2002 the governors of both Georgia and Alabama changed, and this led to major changes in Alabama's negotiating team. With the election of a new president in 2000, the federal commissioner also changed.

New representation brought new positions and proposals. For example, with the change of governors Florida switched from a proposal based on demand limitations for upstream users to one that was based on minimum flow guarantees at the border. This position came as an unpleasant surprise to some Florida stakeholders who had not been consulted. Associated with this decision to change from a demand-based to a flow-based Allocation Formula was an edict by Georgia negotiators that the agreement could either include demand limitations or minimum flow guarantees, but not both. Georgia was never asked to justify this edict, and many downstream stakeholders did not accept the proposition that the two positions were mutually exclusive.

Stakeholders were brought into the process by each of states as they chose to bring them in. Florida initially held meetings every month to educate stakeholders on the issues, to brief them on the status of the negotiations, and to ascertain their interests. As the Allocation Formula negotiations progressed, these meetings became less frequent. Toward the end, Florida stakeholders had to call for meetings themselves. In Georgia, stakeholder involvement was handled through the Governor's Advisory Committee, which did little more than hold meetings at which the Georgia negotiating team briefed stakeholders on the status of the negotiations. In Alabama, stakeholder meetings tended to be large public meetings that focused on the negotiations in both the ACF and an adjacent river basin, the Alabama-Coosa-Tallapoosa (ACT) basin. Because far more of Alabama is in the ACT basin, the meetings tended to spend most of their time on it.

The federal team attempted to assist the states by providing Instream Flow Guidelines for protecting riverine ecosystems and a Conceptual Water Allocation Formula. The state negotiators essentially ignored both.

Two stakeholder groups developed their own proposals in an attempt to break the stalemate, but neither proposal was given much credence by state negotiators. One was the Mid-Chattahoochee Group, an informal group of interests in the middle Chattahoochee River that included representatives from an array of interests in Alabama and Georgia, including the city of Columbus, Tri-Rivers Waterway Development Association, industries such as Mead Paperboard, and the Chattahoochee River Keeper. The Nature Conservancy (TNC) made the other proposal. It had a staff person dedicated primarily to working on the negotiations, and brought in assistance from other parts of the country. Representatives from TNC's Freshwater Initiative introduced the concept of the Natural Flow Paradigm that became one of the foundations of the negotiating position of Florida and conservation groups.

Conservation groups in Georgia, Alabama, and Florida banded together to form the Tri-state Conservation Coalition in August 1999. Eventually more than 45 groups joined. The Coalition's mission was to "unify and coordinate the voices of the regional conservation community in order to advocate for protection of water quality, biodiversity, and recreation in the ACT/ACF River Basins." The Coalition is steered by a provisional committee comprised of several of its members. They were instrumental in getting the commissioners to take questions from the public during the meetings (Holmbeck-Pelham 2003). This coalition provided conservation groups a better informed and coordinated voice in the negotiations.

In the ACF Water Allocation negotiations, the stakeholders' role was much closer to token involvement than to participation. In the end, this left the state negotiators with little public support for their draft formulas, the potential for major confrontations from any formula they proposed, and ultimately the termination of the ACF Compact. A similar compact failed much the same way in the neighboring ACT basin in 2004.

Process Design

Adaptive governance requires a decision-making mechanism that addresses the needs of groups involved. Early in the Allocation Formula negotiations, several key decisions influenced the success of the negotiations:

- not to include an outside mediator;
- not to agree on specific criteria or performance standards that distinguished an acceptable Allocation Formula from an unacceptable one;
- to provide entire proposals to each other instead of outlining the various facets of an agreement and then negotiating each of these facets separately;
- to have the policy negotiators also negotiate complex technical issues; and

- to severely limit the involvement of stakeholders.

As part of the Comprehensive Study process, key participants from the three states and the federal government were provided training in alternative dispute resolution. Nevertheless, when the negotiations on the Allocation Formula were initiated several years later, the states chose not to utilize an outside mediator but to manage the negotiations themselves and to alternate the chair from meeting to meeting. In November 2000, when progress stalled, an outside mediator was brought in for an extended negotiating session. But instead of exploring how to revise the process or why the parties could not reach agreement, they spent several days merely rehashing the differences between the parties (which, it could be argued, the states already understood) and whether they could find common ground. The process did change after this session, but not toward a more collaborative process. Instead, the negotiating teams changed from mostly technical and policy experts to mostly lawyers.

The decision not to agree initially upon what defined a successful agreement also led to problems. Instead of jointly defining criteria for a successful Allocation Formula, the state negotiators chose to judge alternative proposals based on their own internal, unspecified criteria. Thus they wound up debating the acceptability of technical details without agreeing what these details were intended to accomplish, a situation which was not conducive to reaching agreement and left each state guessing what the other states really wanted.

Instead of defining issues of difference and working out common ground, the negotiators initially chose to negotiate entire agreements. Attempting to negotiate an entire document at once made the negotiations far more complex than if they had addressed one facet at a time, made it difficult for stakeholders to understand what separated the three states, and made it easier for negotiators to avoid being accountable for their positions. It also left many secondary issues, nevertheless important to some stakeholders, out of the agenda as the negotiators struggled through the primary issues.

Overall, the negotiators established a pattern of avoiding difficult issues by passing them on. The Interstate Compact itself avoided the specificity required for an Allocation Formula, albeit for good reason. But when hard decisions had to be made in the Allocation Formula negotiations, the parties repeatedly chose to extend the deadline rather than confront the problems that kept them from an agreement. Extensions were not targeted to address specific issues or accomplish specific tasks, but only to continue negotiations for some arbitrary period.

Another complicating factor in the negotiations was the tendency of the negotiators to take the role of both policy and technical negotiators. This sometimes led parties to negotiate issues of which they had at best only a cursory understanding. The negotiators should have defined the policy boundaries of an acceptable agreement and let their technical support staff work together on the details. All of these process decisions hindered the development of trust and collaboration in the negotiation process, both among the negotiating parties and between the negotiating parties and stakeholders.

Alternative meeting formats were tried over time. Initially, negotiation sessions were held in public with sessions rotating between the three states. Curiously, these sessions were set up with the negotiating teams facing the public instead of each other. This ultimately resulted in little progress since the negotiators were performing for the public as much as they were negotiating with each other. Later, they had closed negotiating sessions and reported the results to the public. In the waning days of the Allocation Formula the three governors played a more active role and began negotiating with each other both in private and in public.

At times agreement appeared to be near, but the final concessions never materialized, particularly after different interest groups challenged their own state delegation's position. When state positions changed as a result of this pressure, their commitments to negotiated positions lost credibility.

Interaction among the negotiating parties was sometimes collaborative. The negotiating parties did work together to collect water usage data and develop models of the watershed. At other times interactions were strained as one side accused the other of exaggerating claims or not sharing information. Stakeholder involvement and participation at the commission meetings was limited. Individual stakeholders were allowed to make brief comments at the end of the meetings, but not to ask questions of the negotiators. Formal interaction of stakeholder groups from different states was limited even though some shared interests, but some informal interaction did occur.

Ironically, the three negotiating states tended to focus on making decisions on issues that were not in their jurisdictions while avoiding decisions about issues that were. For instance, the draft Allocation Formula proposals developed by all three states provided specific guidance on how the Federal storage reservoirs should be managed, but made limited commitment to address consumptive withdrawal permits.

Scientific Learning

Several science-related issues came to the forefront during these negotiations:

- defining the physical limit for Atlanta's water withdrawals;
- tabulating the actual amount of Georgia's irrigation requirements;
- explaining the relationship between surface and groundwater in the Flint basin;
- understanding the relationship between flows and the productivity of ecosystems, especially for the Apalachicola River and Bay;
- defining indicators that determined when drought was occurring; and
- utilizing adaptive management to allow the Formula to evolve as more was learned about the basin.

Although the experts had agreed on technical tools and the database for water demands during the earlier Comprehensive Study, little progress was made toward resolving differences in the Allocation Formula negotiations. Some critical issues were put off to the implementation period, such as the indicators that would determine whether the adopted formula was sustaining ecological values.

Considering the complex, technical nature of the negotiations, the limited involvement of the expertise of the state universities is somewhat perplexing. The university communities of all three states house considerable talent in many pertinent fields, including hydrology, water management, dispute resolution, economics, engineering, and ecology, that could have contributed to the process of negotiating an allocation formula. Perhaps the state negotiators worried that allowing universities instead of state employees and consultants to play a major role would have lessened their power and control. Perhaps the lack of funding kept professors from the negotiations, or perhaps they were unwilling to apply their expertise unless they were formally brought into the process.

Parallel classes were taught at the University of Alabama, University of Georgia, University of Florida, and Florida State University in 1999. After the classes were finished all of the participants convened a mock negotiation on adopting an Allocation Formula. This effort, however, had no perceptible influence on the real negotiations. Only one of the states sent a representative of their negotiating team to attend the mock negotiation. After it was completed he cynically commented that their approach "would not work in the real world." Of course, in the end, neither did his. This class was continued at Florida State University; several stakeholders enrolled and acquired a greater understanding of technical details and the process. The class was not continued at any of the other universities.

The Comprehensive Study, completed prior to the negotiations, provided the basis for a growing technical consensus. Modeling tools such as STELLA and HEC-5 were developed for hydrologic evaluation of the entire basin, and data from technical reports were incorporated in the models to provide a common means for the different sides to evaluate a proposed allocation formula. These hydrologic models were based on flow conditions in the basin between 1939 and 1993, and extended through 2001 by the Corps of Engineers as the negotiations dragged on. The Comprehensive Study collected data and developed models collaboratively, and expended effort to seek consensus approval of all products. Outside parties such as the Corps of Engineers Institute for Water Resources initiated and guided the process. However, once the Compact was signed and the states began negotiating an Allocation Formula, the emphasis on collaboration and approval of work products ceased, the process successfully used during the Comprehensive Study was abandoned, and the states switched into a more competitive mode of using scientific information. This switch, however, was not overt, but resulted from the negotiating process set up by the three states.

The states had agreed on data sets and modeling tools, but not on methods for analyzing the model output. Should the entire data set be analyzed? Should data be aggregated into monthly or annual time increments? Model output consisted of 63 years of daily values for multiple flow locations and reservoir elevations: over 23,000 data points for each location. Organizing such large data sets so that they can be understood is not a trivial task and generally is not the type of work that is well done by negotiating teams without strong technical support. No consensus was ever reached on how data should be presented, with each party presenting their data as they saw fit. This in turn did not promote understanding among the negotiating parties. Data organization was complicated because the negotiating parties never reached a consensus over what constituted an acceptable agreement, or the objectives to be assessed with the data. In the end, although the negotiating parties had the capacity to share a vision of the physical consequences of a proposed Allocation Formula proposal, they did not use this capacity. In essence, because of problems with process issues in the negotiations, the states were unable to utilize the results of the Comprehensive Study to reach an agreement on the Allocation Formula.

When Georgia's negotiators changed data that had been developed during the Comprehensive Study, this resulted in a further erosion of trust among the negotiating parties. Georgia insisted that the water demand projections for Metropolitan Atlanta in the Comprehensive Study were understated, unilaterally offered new data based on the contention that their population was growing faster than projected in the Comprehensive Study. This new data was not reviewed like the other supporting data: instead, the other parties were simply informed that this was how Georgia was going to evaluate proposals. Georgia also began using both a factor for dry-year irrigation demands and an amount of irrigation acreage in the Flint basin that differed from those in the Comprehensive Study. Again, these changes were not mutually agreed upon or justified as data were in the Comprehensive Study. Combined, these two modifications led to severe depletions of flow in the Flint basin in model simulations. These depletions led to the lowering of reservoir elevations in the model simulations; as a result Georgia negotiators did not accept Florida's proposed minimum flow values. Although the dry-year multiplier and irrigation acreage values were disputed by Florida and by the U.S. Department of Agriculture, Georgia refused to alter them until the waning days of the negotiations. Ironically, recently completed work done as part of a "Sound Science" initiative in Georgia showed that the values Georgia used were in fact excessive. In retrospect, one state's decision not to insist on a collaborative process of approving all technical information contributed to the erosion of the negotiating process and the ultimate termination of the Compact.

The decisionmaking process also had to deal with the issue of managing the basin during both "normal" times and drought. During the Allocation Formula negotiations, the basin was experiencing its lowest flows since 1939. In 1999 and 2000 average annual flows entering Florida from Jim Woodruff Dam were about 40 percent of what they normally were in the 63-year database. The drought focused the negotiations on allocating water during rare extreme events instead of normal conditions. Although from a hydrologic perspective this was the most severe drought in the period of record (1939 to the present), from a climatic perspective the 1954 to 1958 drought was more severe (Leitman, Dowd, and Pelham, 2003). The extent to which low flows resulted simply from a very extreme event or were exacerbated by increased consumptive withdrawals was not analyzed in the negotiations.

Lack of sufficient knowledge about ecological impacts was arguably at the core of the Allocation Formula conflicts. The negotiations were complicated by Florida's demand that flow at the Florida/Georgia border be defined or as the *flow regime* (i.e., frequency, magnitude, timing, duration, and rates of change of flow) needed to protect the ecosystem, not simply the magnitude of flow. When Florida included minimum outflow values at Jim Woodruff Dam based on a flow regime rather than a simple minimum value, the other parties complained that this was too complicated, even though the approach mimicked current operations and simply replaced values needed for providing a given navigation depth with minimum flow release values.

Downstream interests were frustrated because upstream interests expected them to define the ecological ramifications of conditions that had never been experienced. When downstream interests defaulted to the natural flow position (a flow regime similar to what had occurred historically), upstream interests were frustrated and demanded a more precise definition of downstream water needs. Despite repeated denials by Florida, the other negotiating parties repeatedly interpreted this as Florida demanding historical flows and thereby restricting any growth in water withdrawals. Indeed, Florida did not extend the general concepts of the natural flow paradigm to their specific issues, and explicitly listed those flow regime factors most critical to ecological integrity. Florida noted that the scientific basis was not sufficient for specifying flow regime, and initially requested that adaptive management be a part of the settlement to allow for changes in flow requirements based on designated studies of flow impact on the ecology. Upstream interests, however, expressed reluctance to accept the uncertainty implicit in a truly adaptive allocation formula. Later in the negotiations Florida wanted to both lock in their flow requests and use adaptive management, which were contradictory positions.

Substantial scientific learning occurred during the ACF negotiations. The negotiating parties repeatedly refused to make decisions, citing inadequate scientific information, even though decisions are made with far less information and technical capacity in other river basins. The real problem was not the information or tools available to the negotiation teams, but the lack of a process to use information collaboratively and the lack of trust among the negotiating parties.

Public Learning

Public learning is complicated by the existence of four overlapping audiences involved in adaptive governance: technical/scientific, managerial, elected representatives, and the public at large. In the ACF negotiations, stakeholders' access was complicated by the inherent complexity of the issues being negotiated, insufficient scientific knowledge to answer relevant questions, and the sheer number of stakeholders. The information, tools, and tradeoffs associated with managing water in a nearly 20,000-square-mile basin with varying demographics, multiple-purpose reservoirs, and changeable climate are highly complex, and few stakeholders had the necessary background or access to experts to understand the issues well enough to influence the negotiations. What was needed was not a "one-size-fits-all" process. Instead, options for stakeholder involvement should have been available to fit the varied needs, expertise, and level of involvement of participants. This would have entailed more time and expense, but might have increased public support.

The structure of the ACF negotiations systematically undermined public learning. Forums to educate stakeholders and seek their input did not exist for the most part, although such forums were available in Florida during the

Comprehensive Study. Instead, stakeholder forums were generally designed to justify the negotiating positions of the states.

The negotiators were torn between simultaneously negotiating with the other states and educating their internal stakeholders. Educating stakeholders takes considerable time and effort, and risks leaking negotiating positions to the other states. It was also difficult for staff to simultaneously respond to new positions from the other states and educate stakeholders about the current position, which was changing while they explained it. At some point, stakeholders need to accept that the negotiators understand their basic interests and are representing them. This trust did not exist in the ACF negotiations.

Nevertheless, a significant level of public learning by stakeholders did take place in the period between the initial proposal to reallocate water in Lake Lanier and the collapse of ACF negotiations. Although no agreement could be reached, the general level of discourse for the full range of participants reflected a more sophisticated perspective on how the basin could be managed.

Problem Responsiveness

Meeting the interests of upstream and downstream users on a sustainable basis was arguably the major equity and efficiency issue in the Allocation Formula negotiations. Unfortunately, the focus on geographical boundaries in a "water war" among the three states diverted attention from the issue boundaries and the differences between upstream and downstream users.

Specific equity questions relating to the ACF Water Allocation negotiations included:

- Should withdrawals from the upper Chattahoochee Basin by Atlanta be limited, and at what level?
- Should withdrawals for agricultural irrigation in the Flint basin be limited?
- What should be the balance between use of the federal storage reservoirs to support downstream water needs and retention of water to provide benefits to communities adjacent to the reservoir?
- What level of risk is acceptable in providing flows to support downstream ecosystems in the Apalachicola River and Bay?
- What is the balance between making decisions for political and technical reasons?
- What level of involvement was appropriate for the vast array of stakeholders in the nearly 20,000-square-mile basin?

With regard to equity, one of the mantras of the negotiators was the need to "share the pain," referring to shared adversity in times of drought. Georgia did not translate this mantra into Allocation Formula proposals. Its latest proposal could be summarized as providing no limits on consumptive depletions, managing reservoirs at higher than historical elevations, and letting downstream interests have what is left. This position left many downstream interests questioning what pain was being shared.

Conclusions

The ACF Compact was the first Interstate Water Compact in the United States since the passage of the major federal environmental laws in the 1970s, and the first ever in the Southeastern United States. It presented a major opportunity to manage the ACF basin as a system. But the Allocation Formula negotiations resulted in no agreement and the termination of the ACF Compact, leaving the three states with no formal agreements or implementation mechanism.

The Memorandum of Understanding (MOU) between the three governors in July 2003 led directly to the termination of the Compact. The MOU was negotiated between the governor's offices of the three states to provide the basis for negotiating an Allocation Formula agreement. The principal ACF negotiator for Florida, however, was not involved in developing this MOU and did not even see it until it was presented at the Commission meeting at which it was adopted. The MOU was essentially an endorsement of Georgia's negotiating position, which Florida had rejected numerous times because it violated several of Florida's main negotiating positions.

After the agreement was signed, attempts were made to add stipulations to the Agreement to reaffirm some of the basic tenets of the Florida negotiating position. When Georgia's negotiators sent the MOU to their stakeholders, they did not include Florida's stipulations and contended that they never received them. Georgia then provided an Allocation Formula proposal to Florida that was consistent with the MOU, but unacceptable to Florida negotiators and stakeholders. In response, Florida provided an alternative proposal to Georgia one week before negotiations were to terminate that was consistent with their stipulations, but not consistent with the MOU. Florida negotiators told Georgia negotiators that if they did not accept the terms of their alternative proposal, Florida would not agree to extend the negotiations and the Compact would be terminated. At the final meeting, Alabama and Georgia expressed a desire to extend the negotiations and a disappointment that Florida was not following the terms of the MOU.

The termination of negotiations was inevitable for several reasons:

- A breakdown in trust between Florida and Georgia, compounded when Georgia settled litigation on related issues between the Corps of Engineers and Southeastern Power Users;
- The manner in which the MOU was negotiated with Florida;
- Position-based rather than interest-based negotiations, which created competitive negotiation instead of collaboration;
- Failure to have a neutral party to keep negotiations productive;
- The complexity, high stakes, and unknowns associated with the water allocation formula; and

- Internal conflict between an open process for developing each state's position and a closed process for justifying positions and maintaining an advantage in interstate negotiations.

Despite the termination of the Allocation Formula negotiations, there were several major gains to the citizens of the basin as a result these negotiations. First, the paradigm for many people in the basin has expanded from a local or single-use perspective to a watershed perspective. The public learning throughout the negotiations could produce better management of the basin in the future.

Second, significant information and analytic tools have been compiled to understand the management of water resources in the basin. This may mean more proactive and informed involvement of stakeholders in future decisions.

Third, when faced with either accepting an agreement contrary to their interests or letting an historic basin-wide institutional mechanism end before it began, Florida chose to stand up for the interests of its stakeholders and let the Compact terminate. Given the immense political pressure to reach an agreement, Florida's rejection of an unacceptable agreement can be considered a success.

Finally, there were multiple institutional changes in the three states as a result of the ACF negotiations, including the establishment of an Office of Water Resources in the state of Alabama, the passage of the Flint River Drought Protection Act in Georgia, the creation of the Tri-state Conservation Coalition by environmental and resource-based groups, the creation of the Mid-Chattahoochee Group as a coalition of business, municipal, and conservation interests in the mid-Chattahoochee basin, and the enhanced coordination among state and federal agencies in the basin. The prospect of future litigation may stimulate the collection of more data and the funding of studies to answer critical questions relating to the relationship between flows and the ecosystem. Whether or not a litigation-based process can build on these gains, and whether the citizens of the basin can design a decision process capable of overcoming the problems revealed by the ACF negotiations, remain to be seen.

In the end, an opportunity was missed. Based on my involvement in this process, I contend that an answer existed that was technically feasible, politically viable, and consistent with the interests of all parties. The failure was in the negotiation and political processes.

Notes

1. I wish to acknowledge the helpful comments and contributions of Karla Brandt, Skelly Holmbeck-Pelham, Jerry Ziewitz, and Bill Werick in writing this paper. This chapter was adapted from an earlier version of this case study that was written by the Water Governance Work Group, Florida State University, and Bruce Ritchie.

CHAPTER 7

Everglades Restoration and the South Florida Ecosystem

Michael R. Boswell

In the early 1970s the U.S. Army Corps of Engineers completed a project that had been started nearly 120 years earlier by the citizens of the new State of Florida: taming the wild Everglades for the productive use of society. The Everglades, an enormous wetlands complex that dominates the South Florida ecosystem, was finally under the full hydrologic control of an extensive system of canals, dikes, and pumps. Also by the early 1970s, the population of Florida's lower east coast had reached 2.2 million, up from 694,000 in 1950. This dramatic increase in population and the massive alteration of the Everglades caused numerous problems. Debates erupted over insufficient natural surface water flows to Everglades National Park, altered nutrient flows and exotic species threatened the natural vegetation in the Everglades, and studies showed that the water quality of Lake Okeechobee and Florida Bay was in decline. Moreover, a 1968 U.S. Army Corps of Engineers (USACE) report, *Water Resources for Central and Southern Florida*, concluded that South Florida could not meet projected urban and agricultural water supply needs beyond 1976. The evidence was mounting that the damaged South Florida ecosystem was failing to support the needs of natural and human systems alike.

The government institutions responsible for the South Florida ecosystem began responding to these problems. The history of these responses, from the early 1970s to the present, shows a transition to new governance regimes that embrace some aspects of adaptive governance. This transition has been slow and difficult. The challenges for adaptive governance that are discussed in this volume provide a meaningful analytic lens for examining this transition. I address four of the challenges: public learning, representation, process design, and scientific learning. By way of introduction, I examine the early responses to problems in the Everglades and show that the challenge of public learning was evident at that time—it took about two decades for independent governmental institutions to learn they had a collective problem that required a collective solution.

Incremental Change and the Failure of Governance Institutions

As evidence of problems in the Everglades mounted, the governance institutions that had created the problems began incremental measures in hopes of maintaining the ecosystem's integrity. The first notable attempt was in 1970 when Congress passed legislation (PL 91-282) that guaranteed water flows to Everglades National Park. These flows were set as annual minimum allocations that were "divided into minimum monthly deliveries said to reflect discharge characteristics during the 1940s and 1950s," i.e., before the USACE intervention (Light and Dineen 1994, *66*). Soon after this action, Governor Reuben Askew declared a "water crisis" in South Florida and issued numerous restoration recommendations at the Governor's Conference on Water Management in South Florida. This inspired the Florida Legislature to pass the Florida Water Resources Act, which established state policy aimed at protecting water supply, water quality, and environmental values (Blake 1980).

Despite this flurry of activity during the early 1970s, there was little real change until a crisis occurred in 1983. Unusually heavy rains associated with an El Niño weather system persisted through January and February 1983, requiring heavy, undesirable, and off-season regulatory releases of water to Everglades National Park. By March, an environmental emergency was declared in the park and a seven-point plan for immediate action was developed. (Light and Dineen 1994, *74*)

Governor Bob Graham responded to the crisis by establishing the "Save Our Everglades" program, which endorsed the "Seven-point Plan" and several other projects aimed at Everglades restoration. One of the seven points was the field test of a new water delivery schedule for Everglades National Park (Light and Dineen 1994, *74*). Called the "Flow-Through Plan," this involved allowing an unregulated flow of water into the Park, thus ending the minimum delivery schedule in place since 1970. The failures of this experimental plan, however, quickly brought a second experimental plan in 1985: the "Rainfall Plan." This plan was designed to "restore a more natural hydrologic condition" to the Everglades National Park by basing the "amount and timing of water deliveries...on recent weather conditions (rainfall and evaporation)" upstream from the Park (Neidrauer and Cooper 1989, *i*). This plan appeared to have the desired effect in managing water quantity problems.

Also in 1985, Governor Graham established the Lake Okeechobee Technical Advisory Council (LOTAC-I). A year later, LOTAC-I made recommendations for reducing phosphorous loads to Lake Okeechobee by 40 percent, the first serious attempt to deal with water quality as well as water quantity issues in the Everglades. One of the recommendations, for example, led to initiation of the experimental Everglades Nutrient Removal Project in 1988 (U.S. Army Corps of Engineers 1996, *3–6*). These recommendations were timely because they occurred in the same year as a second crisis in the Everglades: massive algal blooms occurred in Lake Okeechobee, possibly signaling the death of the lake. To make matters worse, the next year significant sea grass die-offs were discovered in Florida Bay—the end of the line for water moving through the Everglades system. This new crisis led to passage of the Surface Water Improvement and Management (SWIM) Act by the state in 1987 and a 1988 U.S. Government lawsuit against the state and the water management district for polluting the Everglades National Park (see Light et al. 1995).

A Crisis Leading to the Emergence of New Governance Systems

Hall (1993, *280*) described the characteristics of a crisis that leads to paradigm change: policy experiments attempt to deal with anomalies, fragmentation of authority occurs as politicians "faced with conflicting opinions from the experts...have to decide whom to regard as authoritative," and the debate enters the "broader political arena" of public debate and electoral politics. By the early 1990s, the governance systems in the South Florida ecosystem exhibited all of these characteristics. Organized around traditional ideas about natural resource management that failed to recognize the challenge of public learning, they were clearly failing. This led to the emergence of sustainable development and ecosystem management, and this emergence is leading to a redefinition of governance in the South Florida ecosystem.

In response to the continuing crisis, Governor Lawton Chiles established the Governor's Commission for a Sustainable South Florida (Governor's Commission) in 1994. The Governor's Commission was created "to assure that a healthy Everglades ecosystem can coexist and be mutually supportive of a sustainable South Florida economy" (Governor's Commission 1995, *4*). After 17 months of meetings by this diverse group, the Governor's Commission produced 110 recommendations for improving the South Florida ecosystem, including water supply, pollution prevention, urban sprawl, employment, and quality of life. They claimed that implementation of these recommendations "will bolster the regional economy, promote quality communities, secure healthy South Florida ecosystems, and assure today's progress is not achieved at tomorrow's expense" (Governor's Commission 1995, *12*).

In addition to the Governor's Commission, the federal government had established the South Florida Ecosystem Restoration Task Force (Task Force) in 1993. The Task Force, initially a federal initiative but quickly expanded to include active participation by many state and regional agencies, was created to guide the restoration of the greater Everglades region based on principles of ecosystem management. This included coordination of "the development of consistent policies, strategies, plans, programs, and priorities for addressing the environmental concerns of the South Florida ecosystem" (U.S. Department of the Interior et al. 1993). The first report issued by the Task Force did not make the bold claims found in the Governor's Commission report; however, the report's tone said "if this effort does not work, then we will lose the Everglades."

The Task Force and the Governor's Commission guided the USACE in preparation of the Central and Southern Florida Project Comprehensive Restudy (C&SF Restudy). This was a full review of the original water management scheme begun by the USACE in 1949 and completed in the early 1970s; the C&SF Restudy culminated in the adoption of the Comprehensive Everglades

Restoration Plan in 1999. Thirty years after various governance institutions began to incrementally and independently pursue fixes to the Everglades problem, a comprehensive and collaborative plan had finally been adopted. The governance institutions demonstrated public learning by recognizing why their initial efforts were not succeeding. They adapted in the early 1990s by developing new governance systems.

In the remaining sections of this chapter, I describe the development of the C&SF Restudy and examine how principles of sustainable development and ecosystem management were incorporated. Specifically, I focus on the three principles relevant for understanding the emergence of new governance systems and the potential for achieving adaptive governance in the South Florida ecosystem:

- Holistic approach to problem definition.
- Inclusive and collaborative approach to decisionmaking.
- Experimental approach to implementation and management.

These principles address several challenges to adaptive governance identified in this volume.

The Emergence of a Holistic Approach to Problem Definition

The principles of sustainable development and ecosystem management describe a holistic approach that defines the object of concern on multiple spatial and temporal scales and identifies the interaction among those scales (see Grumbine 1994; Christensen et al. 1996; McDonald 1996). Plans no longer reductively define the object independent of its setting or based on boundaries meant for another purpose. Instead the object is defined by the nature of the problem and can adapt to new or changing information regarding the problem. Christensen et al. (1996, *669–670*) explained:

> There is no single appropriate scale or time frame for management... Nature has not provided us with a natural system of ecosystem classification or rigid guidelines for boundary demarcation... Recognizing that ecosystem functioning includes inputs, outputs, and cycling of materials and energy, as well as the interactions of organisms, ecosystems scientists define ecosystem boundaries operationally so as to most easily monitor, study, or manipulate these processes.

Vasishth (1996, *11*) expanded the relationship between scale and the object of planning by suggesting that the object of planning changes as the nature of the problem changes (as opposed to being fixed by tradition or administrative/regulatory authority) and that once a particular object is defined, exchanges (of matter, energy, and information) across the physical boundaries must be given full attention.

To confront the challenge of scientific learning, the holistic approach integrates a new constellation of scientists and decisionmakers from different agencies into a common perspective. Agencies must establish boundaries that encompass multiple sectors through an integrated approach and interagency cooperation. Since agencies' missions are often established along sectoral lines (such as health, education, housing, welfare, transportation, and the environment), governance is conducted through cooperative agreements, multi-agency task forces, or reorganized agencies. In addition, governance takes on economic and social issues by including non-traditional agencies and by recruiting and retraining personnel to reflect a wider range of skills.

The C&SF Restudy was structured as a holistic plan; thus it offers insight into the value of holistic planning in overcoming the challenge of scientific learning. The holistic structure of the C&SF was manifested in

- a problem definition based on recognition of diminished ecosystem integrity,
- a scope which encompassed all known variables,
- a full integration of governance through interagency cooperation, and
- an integrated study of human and natural systems.

The problem definition established that "the remaining ecosystems in South Florida no longer exhibit the functions, richness, and spatial extent that defined the pre-drainage systems; these systems will continue to decline unless corrective actions are taken" (USACE 1998, ii). In particular, the identified problems were adverse hydrologic conditions in wetland habitats, deteriorated water quality, inadequate water supply for all sectors, and flooding. Each of these was regarded as affecting natural and human systems (USACE 1998, iii).

The scope defined in the environmental impact statement to address these problems was established by the Water Resources Development Act of 1992. The scope was similar to that in the 1948 Comprehensive Report for the Central and Southern Florida Project (House Document 80-643) in that it focused on flood control and on water supply for agricultural and municipal use. For the C&SF Restudy, however, the scope was expanded to include ecological relationships:

> Section 309(1) of the Water Resources Development Act of 1992 (October 31, 1992) directed the Corps to restudy the Central and Southern Florida Project to determine: "...Whether modifications to the existing project are advisable at the present time due to significantly changed physical, biological, demographic, or economic conditions, with particular reference to modifying the project or its operation for improving the *quality of the environment*, improving *protection of the aquifer*, and improving the integrity, capability, and conservation of urban water supplies affected by the project or its operation. (USACE 1994, 77) [Emphasis added]

The recognized problem of reduced environmental quality was to be directly addressed. Moreover, the "focus" was to be on "Everglades and Florida Bay ecosystem restoration while accommodating other water resource demands" (USACE 1994, 77). In addition, the geographic scope exceeded the bounds of the original C&SF Project. It encompassed "approximately 18,000 square miles from Orlando to Florida Bay with at least 11 major physiographic provinces"—essentially covering all of South Florida (USACE 1998, *1-7*).

The broad physical and sectoral scope necessitated a broad scope for the scientists who would work on the C&SF Restudy. Therefore, it was integrated through interdisciplinary and interagency collaboration.

> A multi-agency, multidisciplinary team [the Restudy Team] was created to develop plans to address the problems identified within the study area. This team included biologists, ecologists, economists, engineers, Geographic Information System specialists, hydrologists, planners, public involvement specialists, and real estate specialists from a number of Federal, State, Tribal, and local government agencies. (USACE 1998, *iv*)

The C&SF Restudy Team was composed of staff from 27 public agencies and institutions, including federal, state, regional, and local representation. In addition to the Restudy Team, the Governor's Commission and the Task Force, which contributed significantly during the planning process, were both composed of interdisciplinary and interagency members (USACE 1998).

The challenge of scientific learning was partially overcome in the South Florida ecosystem by the holistic problem definition. Never before in the history of the South Florida ecosystem had such a diversity of scientific expertise come together to work on the same problem.

An Inclusive and Collaborative Approach to Decisionmaking

The principles of sustainable development and ecosystem management describe decisionmaking as an inclusive and collaborative process (Roseland 1992; Grumbine 1994; McDonald 1996; Maser 1997). Collaboration involves creation of a public participation process where all interested parties are continually involved and share power over decisions. Communication among the public, professional staffs, and decisionmakers moves the debate from an initial state of individually held facts and values to a state of collectively agreed upon goals, objectives, and policies. Grumbine (1994, *33*) described the contrast from previous decisionmaking processes: "management through dialog and cooperation at local and regional levels will be quite different from management imposed bureaucratically." Thus, an inclusive and collaborative approach confronts numerous aspects of the challenges of representation, public learning, and process design in adaptive governance.

For the C&SF Restudy, public participation was broadly inclusive and thus provides insight into the potential for overcoming the challenge of representation. First, opportunities for the public to participate were extensive and "ranged from workshops, focus group meetings, educational and technical briefings, presentations to interested parties, public hearings, and newsletters, to having the results of alternative plan formulation efforts available on a web site for comment back to the Restudy Team." Second, the Restudy Team made a significant effort to inform the public of their opportunities to participate. "The Restudy outreach effort was comprised of media engagement, a public information/awareness program, minority outreach, and environmental education." Third, with nearly 150 individual meetings of various interested groups, it could be assumed that the number of participants was high, although exact counts were not available. Finally, the USACE coordinated a public participation program to "1) inform the public, 2) gather information, 3) identify public concerns, 4) develop consensus, and 5) develop and maintain credibility" (USACE 1998).

The challenge of process design may inevitably place limits on representation, which becomes evident when we look beyond who was represented and consider their degree of influence. For the C&SF Restudy, the influence of the general public was limited because the USACE staff and organized interest groups maintained significant control over the decisionmaking process. Involvement of the general public was limited to "two main functions: to inform the public about the Restudy and to generate their input on key issues and concerns" (USACE 1998, *11-1*). The focus group meetings of interest group representatives had significantly greater influence on the decision process than the general public and were considered "extremely productive" (Appelbaum 1999). These meetings were attended by "approximately 150 community leaders, representing a cross-section of interests including agriculture, the environment, water supply, and urban residents" (USACE 1998, *11-8*).

In addition to the focus groups, the Governor's Commission, a diverse and representative group, played a pivotal role in developing the conceptual plan (Appelbaum 1999). The Governor's Commission shared control of the initial plan formulation with the USACE, and significantly contributed to the development of regional-scale objectives for the C&SF Restudy.

> As a first step toward identifying the additional actions needed to develop the Conceptual Plan for the Restudy, the [Governor's] Commission considered 66 options/ideas formulated from a myriad of Federal, state, and local agencies; interest groups; and other members of the public...Through a series of three workshops, the Commission considered and grouped the options together to form alternative plans ... Facilitated discussion allowed for a systematic review and screening of each option ... The result of this process was a list of 40 preferred options, to be evaluated as modifications to the C&SF Project. (USACE 1998)

These preferred options were then evaluated in the USACE's planning process, which began with modeling implementation scenarios and included participation of "representatives from every concerned agency—federal, state and local—[who] worked closely together with other stakeholders to decide which features would be included in each alternative plan." The referenced stakeholder involvement was the focus group meetings discussed above.

> The [focus group] meetings proved a useful process by allowing the various constituencies to comment on whether the appropriate components to be evaluated had been identified and to note what may have been missed in the initial plan formulation phase... The second round of meetings, which took place from September through December 1997, informed the participants about the progress of the plan formulation

> process... A portion of this second round of meetings showed how the comments and concerns provided at the first round of meetings had been incorporated and/or addressed. (USACE 1998)

These meetings were primarily intended to keep stakeholders informed, but the Restudy Team did work to address their concerns.

The challenge of process design was apparent in the relative power of organized economic interests (such as the sugar and mining industries) and governmental agencies, which used their significant political and jurisdictional clout to represent their interests. Most notable was the continued presence of the USACE as the lead agency for planning and managing in the South Florida ecosystem. Although several changes were made to the typical USACE process—notably the creation of the Task Force—the process was more of a modified governance system than a new one. In discussions with various stakeholders, I found that many questioned whether control of the process by traditional agencies would ultimately be effective. The USACE was partially responsible for creating the hydrologic system, did little to react when problems began to appear, and was ultimately forced into restudying the whole system. In addition to the governmental agencies interests, the C&SF Restudy had the potential to produce big winners and losers, both economically and politically. Questions of relative power and ultimate influence over the C&SF Restudy were never adequately addressed.

Adaptive Environmental Assessment and Management

The principles of ecosystem management describe an experimental approach to assessment and management. Holling (1978), who championed the concept of adaptive environmental assessment and management (or adaptive management), asserted that knowledge is always incomplete and changing. Our approach to solving problems must account for this uncertainty and evolution by generating information feedback (through substantial monitoring programs) and by being flexible. Given this understanding, Holling advocated planning and management by scientific experimentation; Christensen et al. (1996, *670*) explain: "management goals and strategies must be viewed as hypotheses to be tested by research and monitoring programs that compare specific expectations against objective measures of results." Holling argued that actions should be designed in such a way that resources are not irretrievably lost, otherwise the potential for learning and adapting is minimized.

To confront the challenge of scientific learning, adaptive management rejects traditional monitoring and evaluation programs that do little to generate new information. These programs were poorly structured and focused narrowly on a dichotomous measure of success or failure (Holling 1978; Lee 1993). Moreover, since monitoring and evaluation was typically the end of a process, the information generated was ignored. Adaptive management overcomes these problems because it is not the end of a process; instead it is comprehensively integrated into a continuous planning and management process. Adaptive management cannot be partially or incrementally adopted because it is a strategy or framework, not a step or technique.

The C&SF Restudy was structured on an adaptive management philosophy. The inclusion of "contingency plans," acknowledgment of uncertainty, and adoption of an adaptive approach indicated that the preferred alternative was assumed to be experimental and conditional. Further, the monitoring and evaluation program was structured as an "adaptive assessment" as described in the Environmental Impact Statement (EIS):

> This strategy calls for the incremental implementation of plan components, with each increment treated as one "experiment" within a stair-step evolution of experiments, each planned and designed to carry the program one step closer to the ultimate goal of system restoration ... Incremental implementation allows both a testing of the hypotheses (and thus provides an essential means for learning more about ecological cause and effect relationships with much greater certainty than is possible with ecological models), and opportunities to refine plans to more effectively meet overall program objectives. (USACE 1998)

The Restudy established project "packaging" and "sequencing" in order to schedule and prioritize the projects. Part of the criteria was the need to prioritize projects that would forward subsequent projects. "Throughout the project implementation process, system-wide analyses will continue. A feedback loop will be established so that each package is evaluated for its contribution to the Comprehensive Plan and that the Comprehensive Plan is revised as necessary to reflect new information developed during the implementation process" (USACE 1998, *10-11*).

The limitations of scientific learning became evident in the difficulty of fully implementing an adaptive management strategy. A report by the U.S. Fish and Wildlife Service suggested why: "In the Everglades, there may never be enough management flexibility to conduct true experiments. To do so would require the ability to manipulate the system on large enough spatial and temporal scales to test major hypotheses" (U.S. Fish and Wildlife Service 1998, *XIV-1*). Furthermore, "An incremental process is required for the South Florida ecosystem restoration program, because of the large and complex nature of the ecosystem and its problems, and because of the uncertainties regarding the ecological responses that will occur as more natural hydrological conditions are established" (USACE 1998, *10-4*). Thus, instead of true field experimentation, the USACE used a modeling approach to test scenarios and provide feedback into the decisionmaking process. The key to the success of this approach will be the ability "to reach consensus on a strong set of hypotheses that describe causal relationships affecting the altered South Florida ecosystems" (U.S. Fish and Wildlife Service 1998, *XIV-1*) and on the interpretation of monitoring results that will be used to adjust the models. Whether the decisionmakers can develop an operational consensus and maintain a political coalition in the light of new discoveries and crises remains to be seen.

Conclusion

In 1999 the Comprehensive Everglades Restoration Plan (CERP), widely considered to be the largest ecosystem restoration project in human history, was officially adopted. Implementation of the plan is expected to be completed in 2025 at a cost of 7.8 billion dollars, which will be split evenly between the federal government and the State of Florida. In 2000, the federal Water Resources Development Act authorized the first 10 years of funding for the CERP; since then the relevant federal, state, and local agencies have been moving forward with implementation. Although much of the effort has been to fully operationalize the plan elements and assign responsibilities, significant work to physically modify the water control system has begun.

It is clear that the changes begun in the early 1990s by the Task Force and Governor's Commission represented a new way of doing business (Boswell 2000). The collapse of the South Florida ecosystem shifted governance from a focus on a hydrologic problem to be managed by a few specialized institutions to an ecosystem and social system problem holistically governed by citizen-based interest groups and public agencies working cooperatively. Further, the hubris of scientific certainty was rejected in favor of an experimental approach. Sustainable development and ecosystem management had arrived in the South Florida ecosystem, but the large scale of the problem presented challenges for adaptive governance.

Recommendations

The successes of representation did not include individual citizens, especially those who might have latent interests of which they were not aware. Expanding participation is not without cost, and thus may not contribute to the overall improvement of decisionmaking. Therefore, I recommend that instead of expanding participation through more meetings, more information, or more public input, the institutions in South Florida do more to ensure that public values and opinions are objectively measured, and do more to uncover latent interests and ensure that they are represented.

Second, scientific learning was impaired by the uncertainty inherent in a system of this physical scale and complexity. This uncertainty created greater conflict among scientists and thus required greater attention to consensus building than is generally required of technical specialists. Therefore, I recommend that institutions in South Florida acknowledge the limitations of the adaptive management approach. Specifically, they should recognize the values-based decisionmaking that inevitably fills the void left by scientific uncertainty, and establish ongoing, collaborative processes for integrating public values into the process. This would reinforce the need for integration of scientific *and* public learning.

In this chapter I examined the emergence of new governance regimes in the South Florida ecosystem that were inspired by principles of sustainable development and adaptive management. These new regimes exhibit some characteristics of adaptive governance, but also struggle to deal with many of its challenges. We are not likely to know the outcomes of this transformation in governance for some time. It is, however, a work-in-progress that will be shaped by a multitude of internal and external forces. Now that implementation is proceeding—dirt and water are being moved—governance institutions should systematically reflect on the "second-order collective action problems" as defined by Scholz and Stiftel. Already participants in the Comprehensive Everglades Restoration Plan are second-guessing their decisions and saying the process produced too little change. Thus, the C&SF Restudy may stand as only the latest—perhaps more notable and extensive—round of incremental change implemented by the USACE since the 1970s that will produce little ecological improvement.

CHAPTER 8

Ocklawaha River Restoration

The Fate of the Rodman Reservoir

Mellini Sloan

THE RODMAN RESERVOIR, an impoundment on the Ocklawaha River in north central Florida, is a last remnant of the Cross-Florida Barge Canal (CFBC). The canal, conceived in the 1820s, was designed by the U.S. Army Corps of Engineers (USACE) to shorten shipping lanes between the Gulf ports and the Atlantic coast. Opposition to CFBC by Florida's young environmental movement led to a halt in construction of the CFBC in 1971, but decommissioning of the already-constructed Rodman dam and the reservoir behind it has been mired in controversy ever since.

CFBC was to bisect the Florida peninsula, connecting existing natural aquatic systems with a network of canals, dams, and locks. Congress authorized the project during World War II when national security interests in navigation were high. After the war, support waned, but in 1960 presidential candidate John Kennedy promised north Florida constituents that he would push for the project, and Congress appropriated the necessary funds in 1962. Opposition was organized by 1963, but construction began in 1964, with President Lyndon Johnson's groundbreaking remarks that, "the challenge to modern society is to make the resources of nature useful and beneficial" (Blake 1980, *201*).

In 1968, 9000 acres of land behind the Rodman Dam was flooded. Only months later in 1969, Marjorie Carr joined other scientists, attorneys, and activists to form Florida Defenders of the Environment (FDE) expressly to fight the Barge Canal and protect the Ocklawaha. 1969 was also the year Congress passed the National Environmental Policy Act, which required an environmental impact statement. FDE produced its own impact statement for CFBC shortly afterward (FDE 2003a). Then the Florida Game and Fresh Water Fish Commission issued a report questioning the projected fishing and hunting benefits of the project (Blake 1980, *206*), and a federal Council on Environmental Quality report found that the project could hurt unique wildlife. FDE and the Environmental Defense Fund filed legal challenges, and in 1971 a federal court issued an injunction halting construction. In 1976, Florida Governor Reuben Askew asked Congress to de-authorize the project, after which $2.5 million was spent on a federal re-study. Construction lay inactive for over a decade, but supporters managed to forestall a legislative vote on de-authorization for over a decade. In 1990, the Florida legislature finally and unanimously requested federal de-authorization of the project, with Congressional approval finally taking place over President George H.W. Bush's signature in 1991.

With de-authorization of the canal project, lands procured for the project were transferred back to the state and incorporated to form the Cross State Greenbelt for Conservation and Recreation. The newly formed Canal Lands Advisory Committee (CLAC), composed of experts on ecosystem restoration and management from within Florida, was tasked with evaluating long-term management options for the canal system (including Rodman Dam and Reservoir) and its adjacent lands. The CLAC initially recommended restoration of the Ocklawaha system to its pre-canal state; last-minute political maneuvers reworked the language of the recommendation to require that further studies be undertaken (FDE 2003b). In response to actions during the 1993 legislative session, FDEP and the St. John's River Water Management District (SJRWMD) drafted a 20-volume study evaluating costs and benefits of four management alternatives—full retention of the reservoir, partial retention of the reservoir, partial restoration of the river system, and full restoration of the river system (*Fla. Stat.* Chap. 253). Their final recommendations supported restoration of the river; they called for a cessation to the debates over whether or not the Ocklawaha should be restored, declaring that "no further studies are necessary to answer the question...efforts should be directed instead at restoration of the Ocklawaha River" (Galantowicz and Shuman 1994). In 1995, Governor Lawton Chiles followed the agency's recommendation, and directed FDEP "to proceed immediately in applying for permits to restore the Ocklawaha River" and to "begin an orderly and phased drawdown of the Rodman Reservoir" (FDE 2003b).

Despite de-authorization, the Florida legislature has yet to appropriate funds to decommission the Rodman Dam. In 1995 and 1996, the two counties in which the reservoir lies passed resolutions opposing decommissioning, and in the years since, the local legislative delegation has vigorously opposed appropriation of funds for the project. In 1996, FDEP sought clarification from Florida's Attorney General concerning its authority to undertake restoration without further legislation. The Attorney General's office found that earlier legislative action had given FDEP the ability to implement decommissioning (Butterworth 1996), but without legislative appropriation for decommissioning or environmental restoration, neither FDEP nor SJRWMD has been in a position to do so. Meanwhile, required Special Use Permits from the U.S. Forest Service have specified time tables for restoration which have not been met (Jacobs 2002, *3*). Most recently, in 2003, the legislature passed two bills effectively prohibiting any actions to remove the dam and requiring further studies, but both bills failed to obtain Governor Bush's signature (Smith et al. 2003; Pickens and Baxley 2003; Fineout 2003).

So the CFBC, after nearly 200 years of controversy, continues to serve as a touchstone for the interactions of politics, economics, and environmental ethics in the management of water resources in Florida. The fate of the Rodman Reservoir seems unresolvable through existing institutions.

Representation

Views of the Rodman issue reflect two fundamental attitudes about human use of environmental resources (Connelly and Smith 1999). The Rodman reservoir created a new aquatic community which, in turn, led to new recreational users who became supporters for continued maintenance of the reservoir. Save Rodman Reservoir, Inc. has become a leading supporter of retention of the impoundment. They argue that recreation on and near the reservoir is valuable in itself as well as an important economic resource for the region. The two counties underlying the reservoir, Putnam and Marion, have each passed resolutions demanding its preservation, symbolic of the attachment of the local populace to their fishing hole.

On the other side of the battle is FDE, an association of scientists and activists, supported at times by national environmental organizations including the Environmental Defense Fund. These pro-environmental protection groups take a contrasting position, incorporating a broader valuation of the resource. Blake (1980) points out that the environmentalists' objection to the dam stems from scientific understanding of the ecology of the system—its intrinsic and independent value separate from human utility.

The sides have expressed their views through legal challenges, through comments on government studies, in studies they have sponsored or prepared themselves, and actively through lobbying FDEP, the Florida Game and Fresh Water Fish Commission, the Florida legislature, and Congress. On numerous occasions there have been public hearings or workshops at which public testimony has been presented.

Scientific Uncertainty

Policy choices at Rodman are complicated by conflicts over scientific information and assessments of what is "best" for the system and the natural and human communities that respectively dwell within and interact with it.

Some degree of scientific uncertainty is unavoidable when undertaking such an ambitious restoration effort, but the extensive studies undertaken to arrive at CLAC's recommendation have made the majority of involved stakeholders confident in the "partial restoration" alternative. The success of similar efforts of ecosystem restoration within SJRWMD is also encouraging (McGrail et al. 1998). Restoration of a functioning riverine ecosystem will necessarily require breaching of the dam structures (Cooke et al. 1993), and once structural alterations are complete, management of the river system will have to be flexible to achieve long-term ecological goals, echoing the intent of if not the exact words "adaptive management." Restoration ecologists agree that "restoration of a river or other aquatic system requires replacing not only the predisturbance morphology but the hydrologic conditions as well" (NRCCRAE 1992). Plans for partial restoration rely heavily on self-organization of the ecological communities in response to restoration of natural hydrologic conditions, which will involve experienced restoration ecologists. Adaptive management efforts require extensive monitoring to assess progress, with continued success dependent upon monitoring results being incorporated into continually evolving management strategies.

Intense scientific studies have not resolved the conflict. Some scientists find an inherent value in reservoir systems and do not see the necessity of restoring them. They agree with the Save Rodman Reservoir group that the system is functioning as is, providing habitat to lacustrine (as opposed to riverine) species, and further serves as an important economic resource to the local area (SRR 2000).

Process Design

The debate over decommissioning has suffered from a lack of clarity in the delineated responsibility for decisionmaking. This jurisdictional uncertainty has blocked resolution of the conflict by traditional means, and may be the greatest challenge for adaptive governance. In spite of executive office support, the legislature asserted authority for final resolution of the reservoir's fate in the 1993 enabling acts, dictating that "such plan shall not be implemented until state legislation specifically directing implementation of the submitted plan or a modified plan, as recommended, becomes effective" (*Fla. Stat.* Chap. 253). Controversy arose over whether the legislature formally "adopted" the plan put forth by FDEP and SJRWMD, and FDEP Secretary Virginia Wetherell asked the Florida Attorney General's Office to rule on the authority of the agency to undertake restoration efforts without further legislation. Attorney General Bob Butterworth decreed that the initial legislation provided the agency with the authority not only to undertake analyses and make recommendations, but also to implement those recommendations (Butterworth 1996). Although the governor backed the decision to pursue partial restoration, powerful political interests have continued to block legislative appropriation of funds in multiple annual sessions.

Not only is there confusion as to the authority of elements of state government, there are also questions about ownership of the lands beneath the reservoir. Regardless of the legislative standstill, the Department of Environmental Protection (DEP) was required to apply to the U.S. Forest Service (USDAFS) for a Special Use Permit for the continued occupation (by the reservoir's waters) of lands within the Ocala National Forest. Although deauthorization entrusted the lands specifically acquired by the USACE for the canal to the state of Florida, it did not have the authority to grant other federal lands occupied by the project. FDEP had resolved some of its jurisdictional issues by obtaining environmental resource permits (required for alterations to wetlands) and consumptive use permits (required for this project as it would temporarily "consume" water) from SJRWMD in anticipation of funds to support restoration efforts, but it still needed permission from USDAFS to temporarily continue to "occupy" the forest lands. USDAFS, in consideration of the voluminous studies undertaken by SJRWMD and FDEP and the recommended restoration alternative, issued an environmental impact statement allowing for continued occupation of the federal lands, submitting that temporary submergence of the lands was permissible if the state pursued partial restoration with the ultimate result of restoration of the forestlands (USDAFS 2002a). The permit dispatched to FDEP required adherence to a schedule with drawdown activities beginning in late summer 2003 and restoration activities complete by 2006 (USDAFS 2002b). As part of the public comment period, representatives from Save Rodman Reservoir submitted a challenge of FDEP's authority to manage the system and accept the permit (Metcalf 2002), to which USDAFS responded that they had "no reason to believe that the FDEP does not have the delegated authority to 'agree to the terms of the Special Use Permit'" (SUP) (Jacobs 2002, *3*).

Although the Attorney General has asserted that FDEP has authority to proceed with restoration, it lacks the funds to do so. FDEP depends on legislative appropriations for most environmental restoration, and as funds for the restoration of the Ocklawaha are not forthcoming, the agency declined to accept the SUP, leaving the forest with an unauthorized "squatter" tenant. Seemingly frustrated by their own inability to undertake the prescribed restoration, FDEP indicated in the letter declining the permit that "this action [the refusal of permit] opens the door for the U.S. Forest Service to pursue restoration" (Ballard 2002). Under the terms of its Special Use Permits, the Forest Service may choose to proceed with restoration itself in an effort to "evict" the state from its submerged lands by breaching of the dam, and requests reimbursement from the state, as terminated permit holders remain liable for the costs associated with both removal and restoration of the site (USDAFS 2002b). The Forest Service and FDEP seem intent on partnering to move forward with the restoration; however, a funding source has yet to be identified (Ballard 2002; Kearney 2002), leaving the conflict between state and federal parties unresolved.

Public Learning

Public views of the relative merits of economic and preservation uses of natural resources have certainly changed in the 50 years since Congress first authorized the CFBC. These changes have played out on a large stage, and the battle over the CFBC was but a small part. In the 14 years since formal de-authorization of the canal, however, the debates surrounding decommissioning of the Rodman dam have taken place with less widespread notice. The preponderance of interested citizens live within a hundred miles of the dam. Both FDE and Save Rodman Reservoir Inc. have sought to involve a wider public, and on occasion have succeeded in mobilizing other organizations to join them in their campaigns, but in recent years this has been a controversy mostly among locals.

In the activist days of the 1970s, studies about the canal prepared by governmental and environmental organizations were widely distributed and influenced public opinion. In the 1990s, the studies have been less elaborate and have received less extensive distribution. Evidence suggests that those in the public who care about this controversy have cemented their opinions over time; few have adjusted their views in response to emerging science or policy deliberations.

Problem Resolution

Framing the future management of the system within distinct alternatives with charged names like "restoration" and "retention" may have reinforced the perception of mutually exclusive visions for the Ocklawaha River. The 1993 FDEP/SJRWMD study did evaluate a range of scientific and economic information before it recommended partial restoration in 1995. But even the technical thoroughness of the 20-volume study could not address the needs of the members of the Save Rodman Reservoir group; they still love their fishing hole and seemingly cannot accept its removal. Members of this alliance may have perceived its outcome as unacceptable and thereby disregarded the level of effort expended in its conclusion.

Although the fate of the CFBC is certain, contrary notions of resource valuation and scientific assessments, compounded by lukewarm political interest, defaults to the status quo—the dam and its impoundment remain, and funds for restoration have not been authorized. In spite of support from the governor's office, state and regional agencies, and legislative bodies in the early and mid 1990s, local resistance forestalled decisive action. More recently, shifts in the composition of the legislature have somewhat turned the tide against restoration.

The two main stakeholder groups hold mutually exclusive visions for the ecosystem. Both visions embody individual preferences for interaction with the river or reservoir. As the groups perceive their interests to be mutually exclusive, potential for negotiation towards conflict resolution is limited, despite extensive studies to resolve issues of ecologic and economic uncertainty. The bills in the 2003 legislative session reveal a widening fracture between Florida's governor, agencies, and legislative bodies. Governor Bush believed there had been "an agreement [between the governor and the legislature] that there was going to be a détente"—no Rodman legislation would be introduced during the 2003 session (Fineout 2003). A breakdown in such agreements does not bode well, as progress requires consensus among the state's two legislative bodies and the governor's office. Correspondence between state and federal agencies does indicate continued interest in cooperative action.

Perhaps mutually acceptable solutions exist, but have not yet been identified. Resolution of the conflict might be feasible were the geographic scope of the problem expanded—the Ocklawaha basin is blessed with multiple large lacustrine systems. Although plans for restoration of the free-flowing Ocklawaha include recreational use, they differ from those currently associated with the reservoir system. Perhaps paired restoration of the river with increased management of nearby lakes to allow access and support for desired fish species would satisfy the desires of recreational fishermen. The process lacks consensus on project goals and credible commitment by stakeholder groups to resolve the conflict—a willingness on the part of each group to yield some portion of their demands. The inefficiencies of the current system reflect the need to adapt to changing conditions and to respond to shifts in political will. When decisionmaking is not linked to funding, these shifts foster inefficiency: decisions are made but not implemented. Fragmentation of responsibility among the agencies charged with management of water resources makes effective management impossible.

CHAPTER 9

Aquifer Storage and Recovery

Technology and Public Learning

Eberhard Roeder[1]

IN THE FALL OF 2000, a water supply coalition of state resource agencies, and water utilities, with the support of the Secretary of the Florida Department of Environmental Protection (FDEP) and the governor, urged the Florida Legislature to pass a law easing regulations governing Aquifer Storage and Recovery (ASR). Because ASR was an important and expensive element of the Comprehensive Everglades Restoration Plan (CERP), the coalition attempted to justify the new rules in part with cost savings for environmental preservation.

Opposition to this change came from two directions: the U.S. Environmental Protection Agency (EPA) and a number of scientists not associated with regulatory agencies advocated maintenance of the status quo until more specific scientific data, particularly about potential contamination from ASR, were available; environmental groups did not trust the expedited permitting process, and opposed any move that might increase pollution of water resources. Together, they successfully challenged the validity of proponents' claims about the science and environmental benefits of ASR. After increasing public debate and opposition to the proposed law during the 2001 legislative session, proponents first rejected amendments that would have implemented an adaptive management approach to the regulatory change, then finally withdrew the legislation in the final days of the session.

Most case studies in this volume focus on multiple issues affecting a specific watershed, whereas the ASR conflict is unique because it involves a single, very technical issue affecting multiple watersheds across the state. The issue combines concerns of water quantity, quality, and habitat restoration in a surprising way. The case illustrates the problems that arise when the state legislature becomes the primary institution for resolving issues related to emerging technologies. It exemplifies the challenges of representing concerned members of the wider public in meaningful technical debates and of designing a process that can resolve both scientific and public uncertainty. By coincidence rather than by design, the CERP process described by Boswell in Chapter 7 played an important role in supporting scientific and public learning about ASR, complementing the aborted legislative process.

ASR and the Water Resource System

The management of water depends in part on the management of underground space. Scientifically speaking, the water-filled underground space has traditionally been divided into regions that are more permeable and allow water to flow relatively easily (aquifers) and regions that are less permeable and impede flow of water (aquitards) (Freeze and Cherry 1979). The technology of ASR consists of injecting water through a well into an aquifer and subsequently recovering water by pumping out of the same well (Pyne 1995). Essentially, ASR uses underground space as a storage reservoir to save water when surface water is plentiful, for use when it becomes scarce. The boundaries of an ASR system in Florida are a low-permeability aquitard below which water is injected, denser brackish water or another aquitard on top of which the injected water may form a "bubble," and a mixing zone around the well within which water quality is affected. The existence, shape and size of a definite bubble, or storage zone, depend on hydrogeologic factors such as density and permeability contrasts and pumping rates (Merritt 1985; Missimer et al. 2002). Because this information and hence the boundaries of the ASR bubble are subject to different interpretations, the ASR system is a social construct as well as a physical reality (Sundqvist 2002). The public learning surrounding this social construct played a primary role in the legislative process.

Rules of Use

Conditions requiring adaptive governance frequently arise among institutions when each focuses on a specialized aspect of water management. In the case of ASR, two sets of institutions for quality and quantity of groundwater have evolved in parallel that address the collective action problems of appropriation (allocation of yearly water quantity) and provision (maintaining the quality of the resource system)(Ostrom 1990).

Quality of water is regulated by the federal Safe Drinking Water Act (SDWA) (42 *U.S.C.*) and the Underground Injection Control (UIC) Rules (40 *CFR* § 146) established in 1980 by EPA. After deep well injection of waste became widespread as a disposal method in the early 1970s (Freeze and Cherry 1979), these rules evolved in response to concerns about the impact of wastes on increasingly scarce underground sources of drinking water (USDW).

Drinking water standards limit the concentrations in water of some compounds that are common and known to have adverse effects. Primary drinking water standards protect human health, while secondary drinking water standards address aesthetic and operational issues such as color and odor.

FDEP operates the UIC program in Florida, subject to oversight by EPA. Florida's groundwater classification system (62-520.410 *Fla. Admin. Code*) has in effect divided underground space between USDW, which are now or may become useful for obtaining drinking water, and deeper, high-salinity space. Only water meeting drinking water standards can be injected into aquifer space classified as USDW (CFR 40-146 and 62-528 *Fla. Admin. Code*). This applies to ASR wells. Wastewater disposal wells, which are used extensively by municipalities in Florida, inject into deeper space and don't need to meet drinking water standards. Both ASR and wastewater disposal rely on the existence of an aquitard barrier to keep the injected material in place. For wastewater disposal the prevention of vertical fluid movement by an aquitard is also a regulatory requirement (EPA 2000).

Since ASR involves both injection and extraction, operators must meet regulations governing the quantity of water extracted from ASR operations as well as the quality regulations. In Florida, the water management districts have authority to allocate most water withdrawals for public consumption, agriculture, and industry through time-limited consumptive use permits (CUP) (*Fla. Stat.* 373.203–373.250, see Chapter 1). Thus management districts are involved as regulators of extraction from ASR and other groundwater sources.

Unresolved Issues of Water Quality

Contrary to scientific predictions and regulatory requirements, injected wastewater has not remained confined to the bubble of injection in several Florida locations (Ruhl 1999; Bernstein 2000; Salamone 2002). Suzi Ruhl, founder of the public interest law firm Legal Environmental Assistance Foundation (LEAF) and a longstanding opponent of UIC practice (Ruhl 1999; Salamone 2002) has argued that underground movement of wastewater confirms misgivings about the reliability of regulatory science. LEAF, FDEP, and EPA reached a court settlement in 1998 committing EPA and FDEP to enforcement of existing regulations, although implementation was delayed (Ruhl 1999; Sutherland 1999). Instead, EPA proposed a rule change in July of 2000 that would replace the "no fluid movement" requirement for the problematic municipal wastewater disposal wells with either requirements for treatment to drinking water standards or detailed investigations and risk management to avoid exposure of the public (EPA 2000; Johnson 2003). Wastewater utilities protested the increased costs, and have so far continued injections in the absence of new rules (Johnson 2003).

The partial failure of the first-order regulatory institution stems from the immediate monetary savings obtained by operators and providers by not following the rules. The lack of action is justified in part by the lack of apparent health problems. While USDW has been impacted (Ruhl 1999; Paben 2003), public health is not affected as long as water quality remains high at existing supply wells (Hui 1999; Deuerling 2003). These arguments would return in the context of ASR.

Expanding Issues Connecting Quantity and Quality

The growing demands for water described in *Tampa Bay*, *East Central Florida*, and *Everglades* play a critical role in ASR as well. Using the same well to recharge the aquifer and subsequently withdraw water from the aquifer was pioneered in the late 1970s by civil engineer David Pyne and the consulting company that he worked for, CH2M Hill. ASR got its start in Manatee County, Florida in 1983 as a tool for storing seasonal rainwater when it was plentiful and recovering the water during the dry season. It spread to public water systems in the Tampa Bay area and Southeast coast, and has risen in popularity since the late 1980s (Pyne 1995). However, by 1998 geologists from the Florida Geological Survey and Florida State University had discovered significant concentrations of arsenic and uranium, both regulated under primary drinking water standards, in the recovered water from two ASR sites in Southwest Florida (Arthur et al. 2001).

Agricultural interests were also involved in early ASR development. In experiments in the Lake Okeechobee area from 1989 to 1991, the South Florida Water Management District (SFWMD) diverted phosphorus-rich, coliform-bearing canal water that did not meet drinking water quality standards from Lake Okeechobee, injected it, and made it available for irrigation. The experiments recovered less than 40 percent of the injected water, an indication that no well-defined bubble had formed (Pyne 1995; Reese 2002).

In the late 1990s, efforts intensified in southwest Florida to use ASR to store disinfected effluent from wastewater treatment plants (reclaimed water) for irrigation purposes (Ellison 2003). FDEP developed rules for this application of ASR, but required that injected water meet primary drinking water standards (62-610 *Fla. Admin. Code*; York 2003).

Several water utilities considered using untreated surface water or groundwater as a source of injected water (Reese 2002). Coliform bacteria, an indicator of contamination by human feces and subject to a primary drinking water standard, are likely to occur in excessive numbers in these water sources (Pyne 2002). Two avenues for permitting the injection of such water were considered, both requiring the isolation of the ASR bubble from all surrounding underground sources of drinking water.

The first would exempt a USDW or a portion thereof under the assurance that the aquifer space is neither now nor will in the future be used for drinking water. If aquifer space can be used by ASR in the production of drinking water it is by definition not a good candidate for this exemption. The only two exemptions ever approved by FDEP occurred before 1990; one was the experimental ASR well for agricultural irrigation at Lake Okeechobee. EPA approval is required for exemption decisions by FDEP for high-quality, low salinity aquifers (62-528.300 (3) *Fla. Admin. Code*). Interviewees at both FDEP and EPA indicated that their agencies generally oppose exemptions (Deuerling 2003; Harvey 2003).

Alternatively, variances can be granted for a particular drinking water standard in a particular project. Public water supplies and human health must still be protected, but the standard must only be met at the border of the mixing zone surrounding the bubble. FDEP grants these variances commonly for secondary drinking water standards such as color and odor, but only EPA can grant them for primary (federal) drinking water standards (Deuerling 2003).

The Political Context

After unsuccessfully pursuing aquifer exemptions with FDEP and EPA, proponents of using alternative source water for ASR decided to seek a change at the state legislative level to give FDEP the power to waive primary drinking water standards in ASR permits. Such legislation would also prepare for a more general rule change by EPA to allow injection of water that does not meet drinking water standards (Pyne 1995, 2002).

Supporters of a legislative change became active in the capital, Tallahassee, between 1998 and 2001. These included David Pyne, the Florida Section of American Water Works Association (AWWA, a water utility group), and Azurix, an Enron subsidiary interested in privatizing water supplies and using ASR with treated wastewater. ASR's potential for advancing privatization did not concern the respondents in this study, but it may have helped in getting the strong support of Governor Jeb Bush and FDEP Secretary David Struhs for the ASR legislation (Florida Public Service Commission 2001; Mann 2001; Pollick and Davis 2002; Pyne 2003).

Proponents introduced ASR legislation in the 2000 session. FDEP initially opposed the bill, but became supportive after negotiated revisions. The revised bill did not get out of legislative committee, but did motivate a more coordinated effort of ASR proponents, which now included FDEP and the three water-poor water management districts (South, Southwest, and St. Johns River), for the next legislative session (Mansfield 2003).

Up to this point the process was limited to the water utility interest group contacting technical agencies and presumably key executive and legislative leaders. Ruhl's LEAF, with its expertise in underground injection, was the only organized opposition. The Comprehensive Everglades Restoration Plan (CERP) changed that. In order to minimize water distribution conflicts, CERP relied heavily on more than 300 proposed ASR wells located largely around Lake Okeechobee to store untreated wet season stormwater runoff for release during the dry season (Purdum 2002). The ASR role in CERP was proposed in 1998, but only at the end of the 2000 legislative session could state and federal agencies reach agreement on implementation and financing (Pyne 2003). Potential savings from storing only partially treated water were by then estimated to save about 5 percent of the total restoration bill (Florida Senate 2001).

The CERP embrace of ASR changed the dynamics of scientific and public learning. Instead of adapting the human system to a given water quantity, CERP proposed to adaptively engineer the ecosystem to provide water for all uses. Instead of a few water utility staff and consultants addressing permitting issues, large numbers of scientists with a variety of backgrounds began investigating every aspect of ASR. Now ASR was linked to the survival of the Everglades, not just operational benefits for some water utilities. In keeping with CERP's adaptive management approach, EPA suggested that it could use a flexible approach to test injection of water violating the coliform standard during a pilot phase of the CERP implementation (Vecchioli 1999; Banister 2001; Florida Senate 2001; Harvey 2003).

The newly proposed ASR legislation (HB 0705 and SB 0854) would have assigned ASR a specific underground space, in which the primary drinking water standard for coliform bacteria and secondary drinking water standards would be waived. This would allow injection of coliform-containing water without aquifer exemption or variance by EPA or FDEP. As a safeguard, the bill required "risk-based demonstrations" of die-off of coliforms in the bubble and no adverse effects on human health. The risk-based demonstrations left the determination of acceptable levels of risks and uncertainty up to FDEP's permitting process. After construction, operators would follow an FDEP-approved self-monitoring plan to detect violations of the permit; FDEP could order an ASR operator to remedy any problem. No one but the ASR operator would have access to the stored water. For proponents focused on reducing costs, these safeguards were sufficient (Struhs 2001; Kwiatkowski 2003; Pyne 2003). Opponents objected that these were just the same arrangements that had proved ineffective in municipal wastewater injection.

In preparation for the proposed change in Florida's UIC program, FDEP requested EPA and Governor Bush requested President Bush to waive the UIC requirements for injected water to meet drinking water standards (Bush 2001; Jehl 2001). In response, the regional EPA stated that it preferred to treat variances case by case (Banister 2001). Beyond that, EPA attempted to remain neutral in state law changes.

Coincidentally, the National Research Council's Committee on the Restoration of the Greater Everglades Ecosystem (CROGEE) began a review of the Everglades Restoration Plan, including the ASR component, during the fall of 2000. It concluded that several aspects of ASR needed more research. First, the reliability of monitoring the performance of the ASR bubble needed clarification. Second, water quality changes such as arsenic dissolution, microbial die-off, and the suitability of the water for the intended uses needed further study. Finally, CROGEE suggested a regional study to address the cumulative hydrologic impact of CERP (CROGEE 2001). The report was published just as the 2001 legislative session was about to begin. What CERP had done to broaden public learning about the possible value of ASR, the CROGEE report appeared to achieve for scientific uncertainties associated with ASR.

The Legislature as an Arena of Adaptive Governance

Legislative staff analyzed the bills in February and early March, mentioning the CROGEE report and its concerns in passing (Florida Senate 2001). LEAF assembled a coalition of environmental organizations in March that argued against the legislation on public health grounds (LEAF 2001). Participation at the first committee hearing on March 20, 2001 was balanced between state agencies and water utilities for and environmental organizations against.

As the bills were debated in late March and April, press coverage for the first time described the events of the last three months for the general public. Critics from environmental groups emphasized the prudence of not contaminating water supplies given the areas of uncertainty listed in the CROGEE report. Two non-agency scientists spoke out in opposition to the legislation: CROGEE member and retired USGS district chief John Vecchioli had been involved in

CERP, and Miami geology professor Harold Wanless had found failures in regulatory science in a South Florida wastewater injection case. City and county officials were approached by citizens, and sometimes voiced their own concerns about the legislation (Hull 2001b; Solochek 2001). FDEP, the water-poor water management districts, and the Florida Section of AWWA lobbied for the bill but received less media coverage (Mann 2001).

LEAF approached legislators to amend the bill to make it only applicable to Everglades restoration, given that the main stated argument for the bills was decreasing the costs of that project. FDEP and proponents of the bill in the legislature blocked these efforts, which raised doubts about the real purpose of the bill; some legislators drafted amendments that exempted their area from the ASR legislation (S. Ruhl 2003; Paben 2003).

The senate and house debated and voted on the bills between April 9 and 16, 2001. The legislative sponsors now presented the bill as a more general water supply measure, not limited to Everglades restoration. Representative Jerry Maygarden from Pensacola was successful in excluding the North Florida Panhandle area from the legislation in the house version of the bill (Kam 2001). This turned out to be crucial because it required renewed consideration by the senate, which had on April 11 voted against an amendment by Brooksville Senator Ginny Brown-Waite to exclude 25 North and Central Florida counties from the bill (Salinero 2001a).

In a letter to the editor of the *Palm Beach Post* published on April 14 and in a similar letter released jointly with Secretary of Health Robert Brooks on April 20, FDEP Secretary Struhs argued for the legislation as a cost-effective Everglades restoration and water supply measure (Struhs 2001; Struhs and Brooks 2001). Their claims to "sound science" seemed to contradict CROGEE recommendations. They assured readers that permitting would be controlled by a complete scientific evaluation and that aquifer space currently used for drinking water supply (different from USDW) would be protected. Several newspapers ran editorials opposing the legislation or published an opposing statement by Ruhl (Pyne 2003; S. Ruhl 2003).

During the last week of April, a flurry of amendments to the bill was introduced in the legislature in response to the chief criticisms of the bill. Two would have delayed implementation of ASR until scientific studies had ascertained the outcome of biological contamination. Another would have excluded areas without "confining beds above and below the aquifer" (Florida Senate 2001). These amendments would have moved the evaluation of scientific knowledge from the permitting to the legislative arena. Negotiations between senators, FDEP, and environmentalists continued through April 28, 2001 but did not result in agreement (*Miami Herald* 2001a).

Over the following weekend the sponsor of the bill, Senator Ken Pruitt, and Governor Bush decided to pull the bill, ending debate on the senate floor. On April 30 they declared the bill dead (Salinero 2001b). This effectively ended public learning about this issue in the legislative forum. Bills similar to the failed one were introduced into the house and senate for the 2002 session but were quietly dropped from consideration.

A key element in this outcome was the refusal by proponents to accept amendments that limited the scope of the legislation. Some opponents saw it as a success of citizenry against special water and growth interests (*Sarasota Herald-Tribune* 2001, *Miami Herald* 2001b). Concerns about decreasing reelection chances for legislators and the governor may have played a role in dwindling legislative enthusiasm for the bill (Hull 2001a). Additionally, in May of 2001 Enron was in the process of dismantling its water subsidiary, Azurix, which had lobbied for privatization and ASR (Pollick and Davis 2002).

Representation

The decision process took place in the state legislature, the political system's fallback location for working out second-order collective action problems among specialized first-order agencies. While elected representatives might appear to represent the general public, the specialized legislative process ensured that groups with experienced lobbyists initially had the greatest influence. Representation and articulation of viewpoints was largely based on self-selection by interest groups or specialized agencies. The rule change would have benefited water quantity allocators and ASR operators by reducing pre-treatment costs. Water utilities had initiated the legislation and supported it through the lobbyist of the AWWA, as did the water-poor water management districts, in particular SFWMD (Mann 2001). The strong administration support and the proponents' decision to justify the change by ASR's role in the Everglades restoration, a popular environmental project, resulted in a low profile of water utilities in the discussion.

The rule change would have exempted ASR bubbles from a water quality standard intended to protect public health, and delegated permitting to an agency that in the past had erred in the interpretation of scientific data and been weak on enforcement. The water quality providers were split: EPA cautiously opposed the legislation, while FDEP supported it. Non-agency scientists challenged the science behind the proposal. LEAF, with its specialized experience, the Sierra Club, and the Audubon Society assembled an environmental coalition and, helped by CERP's and CROGEE's high visibility, were effective in reaching the public and raising public safety concerns. Thus the discussion that began in the confines of legislative committees expanded with increased participation by the press and wider public, because opponents publicized the debate in terms that raised concerns.

Process Design and Scientific Learning

Three decision settings can reach scientific agreement on ASR issues: the legislative, CERP, and the permitting process. They operate on different time scales, with different participants, objectives, and success.

The legislative context is not very discerning about technical information. Staff analysis for the house and senate relies on the expertise of proponents, and generally reflects their view. Formally, information was introduced through staff reports and committee hearings, and was amplified by media coverage for the general public. Legislators then had limited time, and the general public only the month of April 2001, to decide if they could trust the information from the administration, technical agencies, and proponents. The issue of trust was also invoked by Wanless (2003), who stated that he was "not very big on the 'trust-me' attitude that was being offered." Wanless was struck by the need to explain to legislators that lighter freshwater moves to the top of brackish water, a core element of the bubble concept.

The timing of scientific reports played a crucial role in the legislative context. Just before the beginning of the legislative session CROGEE (2001) released its report summarizing uncertainties and unknowns of ASR as one of the components of CERP. This testimony by independent experts with no obvious material interest in ASR legitimized the concerns of many opponents. In the previous year, a study on the migration of wastewater from disposal wells in the Miami area into USDW had provided evidence that bubbles were migrating despite the safeguards of the UIC program (Bernstein 2000; Wanless 2003). The increasingly close votes, the amendments, and the final pulling of the legislation indicate that the learning process of legislators took some time without reaching a consensus.

The scientific and technical uncertainties of the CERP project, ASR included, were addressed by two groups that cross traditional agency lines, the South Florida Ecosystem Restoration Task Force and CROGEE. Both are advisors to CERP and serve to legitimize decisions (Sundqvist 2002), but are not directly linked to decisionmaking about quality of injected water. The federal/state CERP implementation process allows experimentation and data collection that can be useful in assessing ASR, and provides a promising location for continuation of the debate that could not be concluded in the legislature.

The permitting process, as discussed in the *Fenholloway* case study, involves mainly the applicant and the permitting agency, though sometimes other groups have sufficient resources and standing to intervene. Regulators have to evaluate how existing knowledge of natural systems and existing decision rules apply to a specific project. With the exception of Pyne, none of the respondents for this study believed that the knowledge needed to evaluate the proposed ASR permit requirements was readily available in Florida.

Opponents who were most concerned about scientific uncertainty hoped that the defeat of the legislation had given Florida a chance to ensure that scientific learning could take place before widescale commitments were made for ASR projects (*St. Petersburg Times* 2001; Vecchioli 2003; Wanless 2003). CERP delayed tests outside the laboratory on the fate of microorganisms during ASR, in part as a gesture of good will to keep the process moving forward (Dean 2001; Ellison 2003; Kwiatkowski 2003). Answers to many questions are probably at least ten years away (Salamone 2002; Bahr 2003).

Public Learning

The short time open to public discussion during the Florida legislative session limited opportunities to reach agreements over stakeholder concerns. Instead of enhancing public understanding, the legislative debate about easing ASR rules reinforced mistrust. One regulatory official, SFWMD's Peter Kwiatkowski, noted that ASR has "taken such a bad hit in the media we're concerned whether the technology is acceptable to the public, we need to have the public support." (Salamone 2002). Proponents blamed distortion or misinformation among environmentalists and media (Hull 2001a; *Miami Herald* 2001b; Pyne 2003). Agency staff generally stated that proponents had not made their case sufficiently persuasive.

Considerable variance remains in interpretations of what is necessary to ensure safe operations of ASR. On one end of the spectrum, the public health effects of water after it has been withdrawn and treated are critical. Although ASR could be considered a proven technology when injected waters meet drinking standards, the surprising increase of arsenic in recovered water raises questions even here. Eliminating access by others to ASR space ensures that no actual drinking water supply would be affected, and at the same time allows economical operations. Bubble confinement and efficient recovery issues are reduced to operational challenges. The potential for reduction in costs by using the aquifer space to store water becomes the strongest argument for changing the rules. This approach is reflected in statements by the administration (Bush 2001; Struhs 2001), and people close to well operation (Hui 1999; Pyne 1995, 2002).

In the middle is the view that safety is dependent on the certainty with which interactions between the bubble and the larger aquifer system are known. Representatives of the technical agencies and scientists generally fall into this group, and public concern in general appeared to stem from these misgivings. The adaptive management strategy of CERP illustrates one response to it. Within this large group, proponents of the ASR legislation argued that the permit requirements would ensure safety, while opponents argued that more certainty was needed before establishing a use regime that could turn out to be unsafe. These two positions differ in how large a role FDEP has in the definition of "safety." Under the proposed legislation, site-specific permitting involves only FDEP and the permit applicant. Opponents say further study, wider discussion, and legislation are necessary to establish new rules. Thus, the degree to which FDEP is trusted influences which alternative is preferred.

Finally there is the view that safety is compromised by any injection that changes groundwater. Certainty of bubble confinement and efficient recovery are of critical importance. Experience with municipal wastewater injection has shown that contamination escapes the bubble, and current procedures do not protect groundwater quality because of lack of enforcement and knowledge. From this perspective, standards for injected water need to be more stringent, not less so. This view is expressed mainly by members of environmental groups (Ruhl 1999; Glenn 2003).

Members of each group generally see their own interpretation of the physical reality as science-based; they charge the other sides with disregarding science and being reckless or fear-driven, respectively. There is also a large symbolic component to the debate, framed as a battle between the primacy of economics and growth and the primacy of the precautionary principle. Both rhetorical tools limited the effectiveness of the legislative process in solving a concrete problem.

Summary

The hundreds of ASR wells proposed for CERP provided an opportunity for a coalition between water utilities, water-poor water management districts, and eventually FDEP to change the regulatory framework of underground injections and ASR. The proposed legislation would have provided two direct improvements for water suppliers: ASR operators would have reduced pre-treatment costs and gained exclusive use rights to aquifer space.

Opponents, in particular environmental groups, were able to defeat the legislation by opening the process to public scrutiny. They pointed to several failures of the UIC program in which regulatory agencies were unable to keep injected bubbles from moving. They also drew on the timely report on scientific uncertainties relative to the ASR component of CERP by a committee of the National Research Council (CROGEE). And they linked ASR to public health, which stirred wider public concern than the issue of environmental preservation. The debate about the legislation became polarized over the symbolic issues of economic primacy and public safety. As the political costs of the legislative change increased, proponents abandoned the legislation. They remained strong enough to prevent legislation requiring that adaptive principles govern ASR at the legislative level. The questions raised during the debate are now addressed only as a matter of building technical consensus in CERP. Experiments on ASR, if completed and conclusive, may eventually reduce the scientific uncertainty and allow a better assessment of opportunities for joint gains. While CERP has played an important role in this case study, linking technical dialogue to broader public learning remains a challenge.

Notes

1. I wish to thank the following for assistance and information useful in this case study: J. Bahr, J.B. Cowart, R. Deuerling, D. Ellison, J.S. Glenn, R. Harvey, D. Hendrickson, P. Kwiatkowski, G. Mansfield, J. May, J.Z. Paben, R.D.G. Pyne, S. Ruhl, J. Vecchioli, H. Wanless, D. York.

Part II
Practicioners' Perspectives

CHAPTER 10

Adaptability and Stability

A Manager's Perspective

Donald J. Polmann

A STABLE WATER SUPPLY of stable quality at a stable price—that's the goal. If it's at hand, then why change? If not, we must change, but how? What assurance is there that an adaptive governance process will work? Will the situation get worse before it gets better; can the policymakers accept that pain? Do we understand the entire problem; can we pose a holistic solution? Is this a trend or just a cycle? How long can we wait before we act? What if we act before we have all of the data? What is the consequence of a wrong choice?

Adaptive processes, incremental adjustment, sequential steps; these all facilitate needed change and progress. However, the great benefits achievable through adaptive processes notwithstanding, the prospect of uncertainty through the course of change instills change avoidance in utility managers as well as policymakers. From a water utility's perspective, known and consistent requirements and constraints form a stable foundation for both operations and decision making. Adaptive means and methods can yield key improvements and bring great progress, but only if they remain secured to that stable foundation.

Scarcity as a Backdrop

As we view the Florida landscape from a bird's eye view, and contemplate our natural water resources, we are overwhelmed with a sense of abundance. The notion of scarcity seems either a fabrication to further an agenda, or a sense of lack for a select group at a particular moment. Given abundance, governance is straightforward. Everyone may take or receive as they desire. The challenge of governance, and the need to conceive and apply adaptation, arises when the sense or reality of abundance is lost.

Local scarcity and reliance on extra-territorial water supplies in the Tampa Bay region dates to the early 1930s, when the city of St. Petersburg in Pinellas

County developed a groundwater supply in neighboring Hillsborough County. With local groundwater supplies deteriorating due to encroachment of saline water, St. Petersburg reacted to local scarcity by looking beyond the city and county for reliable supplies. St. Petersburg was a growing population center, among others in Pinellas County, and northwest Hillsborough County was by contrast rural and relatively undeveloped. Today, approximately 80 million gallons per day are routinely produced outside of Pinellas County for delivery to its many municipalities. How has the community, locally and regionally, adapted to that practice and resolved conflicts along the way?

Moving from the 1930s to the early 1970s, the Southwest Florida Water Management District (SWFWMD or District), the U.S. Army Corps of Engineers, and the local governments pursued water management opportunities in water-rich central Pasco County. District reports detailed water resource scenarios involving groundwater production, flood control, and ecosystem management. The Cypress Creek Wellfield was the first major effort of the newly-formed West Coast Regional Water Supply Authority (WCRWSA), a cooperative effort among Pinellas and Pasco counties, St. Petersburg, and SWFWMD. Over time, a complex set of multiparty contracts and operating arrangements under the cooperative structure of the WCRWSA built additional regional wellfields and transmission mains. Still, member governments of the WCRWSA could independently operate their own water systems, which included other wellfields. Integrated management was not yet a reality. Although WCRWSA was governed by majority vote of its board, each member government individually approved contracts that would implement capital projects; member governments could effectively withhold their participation and interrupt project implementation by withholding contract approval or funding. In this sense, the ability of WCRWSA to be proactive on environmental issues or third-party claims of inequity was limited. The collective response to local entities impacted by regional operations did not seem very "adaptive." Stable? Yes. Stubborn? Perhaps. Productive? Not particularly.

Adaptation to Environmental Concerns

In general, the challenge of utility system governance is one of equitable distribution: resolving the battles between those who benefit and those who pay (in whatever manner benefits and costs bear on a given situation). In earlier times, that challenge involved an allocation of water resources among competing users seeking real and tangible benefits often associated with economic value or savings.

In more recent times, the concept of "user" has come to include the environment and the water resource itself, represented by those who speak for the environment. Great difficulties and challenges concerning resource allocation arise when the prevailing sociocultural paradigm shifts to one that includes a reservation or set-aside for the environment. This paradigm shift can lead to instabilities among competing users and conflicts with those who speak for the environment. Water management becomes chaotic as competing parties demand changes in restrictions, requirements, and representation. Chaos continues until a new foundation is laid and all users establish stable positions. When the paradigm shift is painful, public, and lengthy it seems that it must be named—thus our region is familiar with infamous "Water Wars" described in *Tampa Bay*.

Public debate rose and fell, but attention to environmental impacts and relevant resource-reallocation and resource-management alternatives ultimately persisted. Debate at times stalled on seeming trivialities like definitions and distinctions between "overpumped" and "overpermitted." Because of the very different implications for liability, and who benefits and who pays, policymakers and elected officials battled for years at great expense before finally overcoming disagreements and achieving a new stability. The new governance paradigm included the "new" form of use, the environment, which is now a tremendous part of Florida's sense of self and has many champions. Adaptations in governance, both rules and process, have in many instances been driven by the "needs of the environment."

Having evolved through a great struggle from WCRWSA, the successor agency, Tampa Bay Water, became the sole and exclusive provider of wholesale water supplies, with an unequivocal obligation to meet member government demands. Water supply-related concerns and conflicts continue between local governments of the Tampa Bay region, some involving equity issues fueled by differences in circumstance and need between densely populated Pinellas County and the growing areas of Hillsborough and Pasco Counties. However, under the new unified structure more and more issues are debated and resolved at the regional level.

As an example of how this new stable foundation supports adaptation, consider the Partnership Agreement between SWFWMD and Tampa Bay Water to deal with the escalating cost of water supplies. The District set aside a portion of their ad valorem tax funds for a New Water Sources Initiative. In exchange, Tampa Bay Water and its member governments agreed to reduce groundwater use at 11 wellfields to a total less than half what was previously permitted. The District pledged $183 million toward the cost of replacement supplies built by Tampa Bay Water. As such, public tax dollars would be spent in the public interest of environmental restoration to replace the infrastructure previously permitted by SWFWMD and built by the public suppliers; the tax payers would pay to replace what the rate payers had paid to build. The cost-sharing agreement reframed the issue of affordability, greatly facilitating groundwater use reductions and yielding an ecosystem benefit for the general public. "Majority vote" took on real meaning in decisionmaking circles, providing the basis for a new stability.

So, are today's governing institutions capable of resolving conflicts among competing users of water resources? Simply stated, yes. Processes for conflict resolution are now defined. Alternative dispute resolution mechanisms have become common; these include mediation, arbitration, and the like, all well-known in contemporary administrative and judicial fields.

Incremental Learning versus Adaptive Governance: A Question of Scale

Is water resource availability, water supply development, water management, water policy, or some other aspect of "water" the limiting factor in this state's future? When and where in Florida is government prepared to state that there is no more water available for additional reasonable-beneficial uses? Does Local Sources First mean Local Sources Only?

It is important, as we examine these issues, to have clearly in mind the disparity of time scales between the evolutionary processes that affect and control our natural ecosystems and the evolutionary processes of governance systems. And we must know and understand the different implications of an evolving ecosystem compared to a cycling one. The forces of nature are greatly variable, and yet cyclical. Any attempt to adapt governance based on a cycle that is mistaken for a trend is doomed to failure. Unfortunately, water users suffering under the natural drought or flood cycle will inevitably be tempted to use adaptive mechanisms to challenge rules and reallocate their costs at times of crisis. The constraints and needs of Florida's water systems may not justify the potential strains imposed by a radically adaptive legal, administrative, and political structure, as witnessed by the expense and decade-long delays inflicted by the Tampa Bay water wars.

It is particularly useful to acknowledge the planning horizon and implementation time frames required for public and private investment decisionmaking for beneficial exploitation of our natural resources. Rules and requirements must be stable over the useful life of a project to allow meaningful, reliable, rational decisionmaking. In the unstable circumstances that may be engendered by any misapplication of adaptive governance, public and private decisionmaking regarding natural resources may become short-sighted and self-serving to avoid commitment of energy and funds to a great unknown. For adaptive management to win the day, adaptive governance must focus on process, strategies, and methods—not on policies, rules, and requirements. If competing users cannot agree upon, know, and embrace the rules, then outcomes will be unsatisfactory at least to some, and the stability needed for long-term resource management will never be achieved.

Incremental, iterative learning in optimizing water resource and environmental management seems to be a natural outgrowth of our scientific and technological advances in data collection, analysis, evaluation, and interpretation. However, embracing this evolutionary process as a method of governance must be approached with caution to avoid the trap of a costly solution looking for a problem.

The Case for Persistence and Stability

Persistence and stability of rules, requirements, and governance provides predictability and fosters effective action. Each major development of water supplies in the Tampa Bay region relied on such conditions. Most certainly, for any number of such outcomes, all players were not equally affected nor were they equally effective in reaching their goals; any concern for disparity or inequity notwithstanding, actions rested on firm foundations. By contrast, periods exhibiting flux in rules, unclear requirements, or uncertainty of governance yielded little progress in infrastructure development or conflict resolution. Such was the case in the 1980s and 1990s when development of the Cypress Bridge Wellfield and the originally planned loop transmission system languished; no new supplies were brought online for a decade while rules and relationships shifted, even as population grew unabated. Similarly, the years of arguments, legal challenges, and negotiations that transformed WCRWSA (a cooperative agency) into Tampa Bay Water (a regional wholesale utility) saw little progress in public service facilities or functionality amid repeated threats to dissolve the agency, even as many stakeholders and the environment suffered. Only a firm foundation of new rules, contracts, and performance requirements finally made possible the conclusion of legal cases and implementation of projects. Thus the governance adaptation process was not a time of doing; it was a time of waiting. Not to be misunderstood, such transformation is certainly a period of great expenditures of energy and effort—great outcomes come only in the final moment.

Restating the premise, if parties to water resource issues expect to succeed through adaptive processes involving large-scale projects, the rules and requirements must be stable, well known, and common to all. Then, with an end result clearly in mind, strategies and procedures—means and methods—guided by those rules can be as adaptive as the players are creative.

Stability and certainty support good investment decisions, both public and private. In the extreme, facilities and infrastructure viewed as "one day an asset, the next day a liability" are unworkable and no such investment would be made rationally. Great difficulties arise when the perception of a neighbor as an asset shifts to that of a liability. Such was the WCRWSA experience, especially with the Cypress Creek Wellfield but also with certain others, where neighbors in dry times cry for you to leave and in wet times plead for you to stay.

If natural cycles induce this shift in perceptions between asset and liability, then implementation changes based on incremental learning within existing rules may be the tool of choice. If a persistent and certain natural trend induces the shift in perceptions, then adaptations to governance may be necessary. While either process can be long and arduous in setup, the outcomes are very different indeed; appropriate expectations of both process and outcome must be established before the proper tool can be selected and the work begun.

In particular, the foundation of rules and requirements must be properly framed in a definable time scale, so that the useful life of a project is examined rationally as a step in decisionmaking. Commitments of energy and funds will be most effective and efficient when the return on each investment can be gauged over the full planning horizon, rather than being clouded by the uncertainty of changing policy goals; the latter leads to an unwillingness to commit into an unclear future, severely restricting the range of options that might otherwise avert short-sighted and self-serving decisions. Incremental adjustments in implementation are naturally suited to temporal adjustments. As long as the framework of rules and procedures remains stable, management can be refined as the future comes into focus.

In times requiring adaptations in governance, stability cannot be guaranteed for the kind of useful project life involved in most major investments, and the challenge here is to get about the process of change in an ordered and productive fashion that minimizes the disruption of vital public service. Even as adaptive governance is effectuated, conditions must be stable until the chapter is closed and the shift complete. Substantive pursuits are possible only after the transformation.

CHAPTER 11

The Power of the Status Quo

Richard Hamann

ADAPTIVE GOVERNANCE requires governing institutions to learn from science and experience and make necessary changes in policy. This chapter considers whether the legal, administrative, and political institutions of Florida have that capability. In the case studies, the challenges of water management have compelled them to learn and adapt. Whether they can change quickly enough, however, is an open question. The constraints on change are substantial. In the absence of natural disaster, the most effective agent for change appears to be litigation.

Institutional Learning

The case studies demonstrate that our water management institutions generally have the capability to learn from science and experience. Political leaders in East Central Florida, led by the St. Johns River Water Management District (SJRWMD), are working to avoid the kinds of conflict and delay in solving water supply problems that Tampa Bay experienced. The Tampa Bay region itself has demonstrated learning and adaptation; it has recognized the harmful effects of declining water levels and the necessity of regional water supply cooperation. By contributing to the formation of a new regional water supply agreement, the Southwest Florida Water Management District (SWFWMD) has advanced water supply and environmental restoration. The Suwannee faced a different threat, a huge increase in nutrient pollution contributed via groundwater and springs discharges. Even without a clear mandate to restore water quality, the Suwannee River Water Management District (SRWMD) has taken the lead in mobilizing diverse resources to subsidize agricultural source reductions. In South Florida, the Comprehensive Everglades Restoration Program (CERP) is an outright acknowledgment by the South Florida Water Management District (SFWMD) and the U.S. Army Corps of Engineers (USACE) that the system of dikes, canals, pumps, and reservoirs they created is fatally flawed and must be replumbed and managed differently. In the *Fenholloway Initiative* we find a clear recognition that siting industrial facilities with massive wastewater discharges on relatively small streams was a mistake that should never be repeated. The parties agree that chlorine should be minimized or eliminated from the pulp-making process, but not how much change is technically and economically possible. Even the restoration of the Ocklawaha River demonstrates some learning. In the 1960s, every institution at the state or federal level supported building a Cross Florida Barge Canal; today even the opponents of restoration concede it should never have been built. Several million dollars in scientific investigations and engineering studies have demonstrated the feasibility and environmental benefits of restoring the river, and every state or federal resource protection agency, including the Governor and Cabinet, supports it.

The Cost of Change

Although learning and adaptation are occurring, it is uncertain that they are coming quickly enough. As the Ocklawaha illustrates, the constraints on change can seem insurmountable, even when the economic stakes are relatively minor. A few powerful legislators, responding to the interests of some local recreational fishing interests, have successfully blocked restoration for decades.

When major economic interests may be affected by change, overcoming the barriers to change often requires substantial public investment. To create Tampa Bay Water, SWFWMD had to commit $325 million of ad valorem tax revenue to subsidize alternative water supplies and water conservation. SJRWMD is offering many millions of dollars to subsidize water supply development in East Central Florida and seeking additional federal funds. There has been a strong push to shift resources from the water management districts and land acquisition programs to subsidized water supply development. Half of the water management districts' share of state land acquisition funds now goes to water supply development, and by 2007 the districts will have spent $1.4 billion.

The farmers in the Suwannee River Partnership are receiving up to 75 percent of the cost of reducing their pollution. Most of the cost of pollution control in the Everglades is being paid for by taxpayers, not the agricultural sources. The cost of replumbing South Florida to reverse the effects of overdrainage and enhance water supplies will be paid not by the beneficiaries of drainage and new water supplies, but by the greater public. In each of these cases, without the infusion of cash from the deeper public purse, those who benefit most from the status quo are likely to have the power, and they certainly have the incentive, to block needed change. Collaboration and cooperation carries a substantial public cost in these cases.

One exception to the necessity of subsidizing change stands out: the Fenholloway River. In this case the owners of an industrial facility are subject to powerful regulatory controls. Under the Clean Water Act, discharge permits must be renewed every five years, and a variety of provisions pressure facilities to reduce discharges and meet water quality standards. Under this regulatory scheme, polluters have limited rights to continue causing environmental degradation. Just as importantly, citizens have clear rights to seek enforcement. As a result of citizen pressure and the involvement of both federal and state regulators, the plants have invested hundreds of millions of dollars in improving the quality of their wastewater. Without the recurring leverage of permit renewal, involvement by both state and federal agencies, a strong burden of proof on the polluter, and effective citizen participation, the improvements would likely never have been made.

Constraints on Change

The legal regime for water resources management can substantially constrain change. Although consumptive use permits must be periodically renewed, in 1997 the legislature required the districts to issue them for 20-year periods. Such a long interval delays opportunities to seek change in sources of supply, mitigation of environmental impacts, and other conditions of water use. As the pulp mill permitting case studies demonstrate, five-year permits can be a much more effective mechanism for change.

There are other constraints on change in Florida's water management system. One of the primary tools for protecting water for the environment is to establish minimum flows and levels (MFLs) (*Fla. Stat.* §373.042, .0421). Although these were made mandatory in 1972, water managers are only now adopting MFLs, after citizen lawsuits compelled them to do so. Meanwhile, the legislature has changed the rules. Where MFLs are not being met, managers are required to adopt recovery plans. Those plans, however, must provide for continuation of existing uses and for new uses. Water suppliers argue that the public should pay the cost of supplying water to offset limits on increased withdrawals attributable to MFLs. In the Everglades restoration, the "savings clause" in federal legislation to authorize the project is being interpreted to guarantee water users existing and even increased withdrawals from the Everglades system *after* the expiration of permits. In both cases water users have managed to secure a priority over environmental needs.

At least with water pollution discharge and consumptive use permitting there is an opportunity to revisit the conditions of regulation. Many forms of environmental regulation, however, allow landowners to continue activities that damage water resources and not even consider ways to reduce those impacts. The owners of drainage and stormwater systems are generally entitled under Florida law to continue operating as permitted. One notable exception that may lead to significant improvements is the federally-mandated requirement for certain stormwater systems to obtain discharge permits. That requirement has arguably been in effect since 1972, but it is just now being enforced. The administrative and congressional delay in implementing stormwater permitting is further testimony to the power of the status quo. A similar evaluation of existing systems is not under consideration at all at the state level.

On the land use side of the ledger, which is inextricably linked to the quality and quantity of water resources, the expectations of landowners in development enjoy a great deal of protection from change in Florida. Rights may be "vested" in ancient, poorly designed developments with substantial water resource impacts. Agricultural land use, one of the most significant source of impacts, cannot be regulated by local governments, and both state and federal control are severely weakened by exemptions. New land use regulation must provide compensation if it creates an "inordinate burden" on property owners or constitutes a "taking."

Under such constraints, change occurs only with great difficulty, when problems have reached such a level of perceived crisis that legislative bodies are obliged to invest funds to "purchase" the right to make changes or suffer the political consequences of changing the rules. Given the magnitude of investment in existing patterns of growth and development, the costs of change are large and there is substantial inertia in the relevant institutions.

The Case for Litigation

It may be that only a crisis can induce existing institutions to accept change or renewal (Gunderson et al. 1995). Sometimes the crisis is a natural disaster or environmental collapse. Widespread flooding was the impetus for draining and impounding the Everglades. Drought led to comprehensive water management throughout the state. Federal water quality regulation was adopted because water quality was declining nationwide, epitomized by a burning Cuyahoga River. The framework for regulating less dramatic sources of pollution, such as pulp mills, was thus created.

Sometimes there must be a crisis not just to adopt legislation and create responsible institutions, but to force the relevant agencies to implement the law. That crisis often takes the form of litigation. Restoration of the Everglades did not begin until the state of Florida had been sued by the U.S. Attorney (John 1994). Tampa Bay Water was created to resolve litigation between the SWFWMD, which had finally begun to enforce rules protecting the environment from over-withdrawals of water, and a host of water users (Rand 2003). The efforts of SJRWMD to negotiate regional water supply solutions in North Central Florida are clearly motivated in part by the desire to avoid litigation. On the Fenholloway River, public interest litigants were able to prod federal and state agencies to be more forceful in demanding pollution reduction. Citizen enforcement is a powerful and necessary tool for resolving environmental conflicts. (Cronin and Kennedy 1997; Environmental Law Institute 1984)

The failure in two of the case studies to make substantial progress toward resolving the conflicts may be explained in part by the lack of opportunities to seek judicial enforcement of clearly defined legal rights. The interstate allocation of water from the *Apalachicola* basin could be determined by litigation before the Supreme Court, assuming that one of the states chooses to initiate the lawsuit. Georgia has no apparent reason to seek judicial resolution; as negotiations have progressed it has continued to allow the withdrawal of ever larger quantities of water, and knows those uses will likely be protected in any future determination. It would likely not be negotiating at all if Alabama and Florida had not blocked action by USACE to change regulation schedules on impoundments. Florida appears reluctant to initiate litigation. Certainly the prospects of securing water for Florida oysters from a highly conservative, development-friendly court do not appear bright. Citizens have no right to bring a case for interstate water allocation; they can only contest isolated permits and decisions. Only when actions of that nature significantly threaten development in Georgia will negotiations make progress.

Ocklawaha restoration is similarly gridlocked. If the dam were being proposed for construction, it could be blocked by citizen lawsuits for violating numerous environmental statutes. Indeed, construction of the Cross Florida Barge Canal was halted when environmental groups challenged the failure to comply with the new National Environmental Policy Act. But with the dam and impoundment in place, there is no need to seek new permits or renewals, and thus limited opportunity to challenge agency action. Although the impoundment occupies part of the Ocala National Forest, and does so without a current permit from the U.S. Forest Service, there is no clear right for citizens to pursue that violation of law, and there has been no political will for the federal government to enforce the law. Consequently, inertia prevails and restoration is stalled. In the absence of litigation, the status quo continues.

Conclusion

One of the principle challenges of adaptive governance is responsiveness to the problems identified through scientific and public learning. In many cases, that requires re-allocation of resources or decisionmaking authority. Those individuals, communities, and institutions that benefit from the current allocation or perceive they will suffer from a change have great power to defend the status quo. Sometimes that power stems from constitutional and statutory protection of certain rights; in other cases it stems from a greater capacity to mobilize political opposition than those whose interests are more remote or diffuse. Even where the law mandates responsiveness to a problem, there may be no action unless interested parties seek to compel it. Litigation cannot change fundamental legal obligations, but it can enforce them, thus overcoming the reluctance of institutions to alter the status quo.

CHAPTER 12

Representation, Scientific Learning, and the Public Interest

B. Suzi Ruhl

As BENJAMIN FRANKLIN astutely observed, "we will know the wealth of water when the well runs dry." In Florida, that day is rapidly approaching because of both quality and quantity concerns. Although water demand has become the issue *du jour*, the discussion has been palliative at best. Root causes of Florida's water dilemma are ignored while clever diatribes skew the debate to mask difficult choices that must be made.

As noted in the Introduction, Florida's current water problems are due in large part to the growth of first-order institutions that enabled their clientele to use water resources without regard to the long-term consequences of such use. Their myopic specialization has too often blinded them to the external effects of their decisions. This myopia poses an increasing threat to human and natural systems, which depend on water for life. Adaptive governance offers to overcome this specialized myopia and direct the necessary attention to the problems that face the totality of water users sharing the same watershed. This essay addresses two primary challenges that impede efforts to develop and implement adaptive governance for decisionmaking about Florida's water conflicts. These challenges are representation and scientific learning within the structure of myopic agencies.

Representation

Myopic agencies thwart effective decisionmaking by impeding representation of relevant interests and issues. Water governance decisions are made in closed systems of agencies and experts who consider only their clientele. Other users are excluded from participation for a number of reasons:

- The agency's attitude may be restrictive.

- The participants may suffer from a lack of knowledge of their true interests and reasonable options.
- Representation may be thwarted by the collective action problem of organizing to facilitate participation.
- Certain participants, primarily those representing the community interest, may lack technical and legal expertise.

The *Aquifer Storage and Recovery* (ASR) conflict illustrates the value of proper representation in decisionmaking and the consequence of failure to include essential parties and issues.

Perhaps the best decisions are those that are made with full participation by all affected parties or their representatives, beginning at the start of decision processes. Consider the analogy between participation and a "table."

Setting the table. All relevant parties should be able to create a venue in which they can raise problems and identify solutions. Too frequently, only those with significant political or financial power can do so. This obstacle to good governance is due in large part to the attitude of the authorities. For water conflict in Florida, the table is usually set only after the menu has already been determined by the government or private sector. Affirmative prevention measures proposed by community organizations rarely receive attention from government decisionmakers and are usually summarily dismissed, which potentially undermines protection of water resources.

ASR, a type of underground storage involving injection into or above underground sources of drinking water, illustrates this. For over a decade, the opponents of legislation to waive the drinking water standard for coliform bacteria for ASR wells had warned of the adverse effects of underground injection. Despite numerous lawsuits, systemic concerns (e.g., lack of an effective confining zone between drinking water and wastewater) were ignored. Yet when private interests proposed legislation that arguably weakened protection, government agencies quickly fell into step in support of it. Public interest advocates were not included in deliberations, and could be heard only when they took their opposition to the media and the public.

Getting to the table. All relevant parties should take part in the debate. Failing to engage those who bear the consequences of exposure to polluted water remains a fundamental flaw in Florida's water governance. This flaw relates to both the scope of representation and the authority of representatives to speak on behalf of particular interests. Further, their interests are not always properly represented by those claiming to represent the environmental position. If parties affected by decisions are involved, it is usually not from the start, and they usually do not have equal access to technical and legal expertise.

Again, the ASR case study illustrates the value of including those who are directly affected in the decisionmaking process. When the legislation was initially proposed, several mainstream environmental organizations did not oppose it, and it moved rapidly through committee. It was only after the issue was presented to the "grassroots" that the momentum shifted the other way. The grassroots included many community organizations and individuals who were on private wells not monitored for contamination or treated pursuant to primary and secondary drinking water standards. Those whose health may be directly influenced by a decision are essential to the process; failure to include them undermines the effectiveness and credibility of the process.

Progressing at the table. All participants must be able to affect decisions. Too often, good governance is impeded by participants' lack of knowledge of their true interests and reasonable options. This encompasses technical and legal issues, core values, and systemic and symptomatic problems. Necessary information that is available is not presented to the broad array of interests, but rather sequestered or ignored. Effective decisionmaking demands that those involved are capable of crafting solutions that should work theoretically and will work practically.

In the ASR conflict, the debate was framed by a limited scope (waiving the coliform standard for ASR) and a particular self-interest (saving money for utilities and government). If the debate had considered short- and long-term concerns, with representation of all relevant stakeholders, then systemic issues (e.g., water infrastructure, supply and demand) could have been raised (EPA 2003). Consideration of more issues would have provided greater opportunity for resolution of the core challenges facing Florida water resources.

The ASR conflict provided an interesting twist on access to legal and technical information. Contrary to usual circumstances, the position of the grassroots was enhanced through the availability of legal assistance (e.g., Legal Environmental Assistance Foundation, Inc.) and technical expertise (e.g., John Vecchioli, P.G.). This contributed to the success of the legislation's opponents.

Ultimately, the resounding defeat by community organizations of legislation that had the full backing of both the private sector and government should provide incentive for effective representation in water governance, which could have led to a compromise solution rather than a defeat for one side that postponed rather than resolved the conflict.

Scientific Learning

Myopic agencies can significantly impede scientific learning. Given the complexity of issues involved in water governance, the high degree of scientific uncertainty due to lack of field data, professional and ideological disagreements, and biased efforts to control the research agenda, it is essential that agencies be vigilant in their approach to scientific learning. To illustrate this challenge, the ASR conflict is evaluated in the context of homeland security.

The massive attention to Florida's water supply overlooks the major concerns of homeland security. Deliberations do not address the vulnerability of critical water infrastructure or opportunities to simultaneously resolve conventional and security risks. Protecting drinking water involves technologies to detect and treat pathogens and other future emerging contaminants. These technologies apply whether the contamination is intentional, accidental, or negligent. They include molecular fingerprints, waterborne genomics, and pathogen identification. Homeland security advances in the public health sector include improvements in syndromic surveillance, expansion of public health authorities at the state level, engagement of health care practitioners in waterborne disease response, and interaction between local health departments and public water supply utilities (Environmental Law Institute 2003).

The ASR conflict in Florida could benefit from this homeland security-derived scientific learning, but has not. A major ASR controversy centered on the fate of microbial contaminants in groundwater. Proponents routinely asserted that microbial contaminants die off in the subsurface, citing salinity and temperature factors. Legislation was proposed to waive the primary drinking water standard for coliform under the assumption that no humans would be exposed to harmful contaminants. But the assumption that all microbial contaminants die, including bacteria, viruses, and protozoa, has not been verified. A major opportunity to clarify the fate of microbial contaminants in ground water arises from emerging technologies, many of which are being enhanced for homeland security purposes, could settle the question. Until the determination is made using technologies that avoid false negatives, the risk to Florida's human and ecosystem health remains. Since the biological knowledge relevant to homeland security is relevant for assessing biological effects from other ecosystem factors on human health, homeland security provides an opportunity to expand the knowledge base for water supply, reduce scientific uncertainty, and improve the rules governing water use.

Another controversy concerned treatment of groundwater. Proponents of the legislation asserted that treatment of the water prior to injection could be eliminated because it would be treated before consumption. Opponents argued that this would put at risk the significant number of Florida residents who consume water from private wells. Advances in the public health sector (e.g., syndromic surveillance) inspired by homeland security could allow adaptive governance to resolve this disagreement.

Finally, the controversy over control of the research agenda could be abated through integration of homeland security. A major concern of opponents of the ASR legislation was the control in the hands of proponents. The research to track contaminants in the aquifer was being conducted in large part by the water management districts and private consultants who proposed the legislation. The appearance of impropriety and bias made resolution impossible. Perhaps the greatest loss from the failure to integrate homeland security approaches into Florida's water debates is the chance to access unbiased technical and health expertise. This expertise is being developed for the core concern of safe water, and is not tainted by self-interest. To the extent that experts and expertise drawn from homeland security can be incorporated into Florida's water debates, many questions can be answered, additional layers of protection provided, and the credibility of the research agenda enhanced.

The failure to incorporate knowledge from homeland security into the theory and practice of water governance epitomizes the challenge of scientific learning. Because the new decisionmaking on drinking water security has not been incorporated into conventional water governance, uncertainties about critical issues remain, professional and ideological disagreements persist, the research agenda remains shrouded in doubt, and opportunities to strengthen protection of this vital resource are lost.

Solutions and Tools

The myopic nature of agencies undermines both representation and scientific learning. Timely, cost-effective decisions that respond to core rather than palliative problems are needed. To make the specialized system accountable, it is necessary to expand participation of affected users of water. It is useful to think in terms of concentric circles, with those directly involved in the decisionmaking at the center. The next circle includes those who may be indirectly affected but who do not have a convenient venue to participate. The third circle can be the general public and natural systems, whose survival is dependent on a healthy water supply. Opening up the current tight-linked specialized decision processes to include these wider circles makes governance of water conflicts more responsive and accountable.

Scientific learning must also take in a wider view. Too often, decisions are based on obsolete technologies and limited consideration of issues outside traditional agency concerns. Given the importance of decisions regarding water, there is no choice but to rapidly evolve the process so that it takes advantage of translational research, public health surveillance, and other emerging tools.

Opportunities to improve governance of water conflicts and thereby strengthen protection of this vital resource abound, and are only limited by the lack of will and creativity of decisionmakers. These opportunities apply to both participation and the breadth of issues addressed. A variety of tools are available to enlarge participation, including stakeholder dialogues, community workshops, participation by decisionmakers in community meetings, videos, educational materials, and curriculum development. Ultimately, education of stakeholders on issues and interests of differing viewpoints can enhance the likelihood of success.

Conclusion

Evaluation of the decisionmaking processes involved in Florida's water conflicts reveals major flaws due to the challenges of representation and scientific learning. Water is consumed at a rate that puts natural systems at risk; responses to water quantity problems increase water quality risks; accountability is replaced with flexibility; and human health impacts of contaminated water are ignored.

Florida's future depends on whether deliberation is based on unbiased assessment of the facts and issues, including admission of scientific and legal uncertainty. It is strengthened through full participation of all stakeholders, especially the individuals whose health and well-being is at risk. In a period of finite supplies, not all issues can be amicably resolved. Yet efforts to use the principles of adaptive governance will narrow the scope of disagreement, mitigate adverse impacts on all parties, and ultimately increase the chance for survival of all human and natural systems.

CHAPTER 13

Adaptive Challenges Facing Agriculture

Martha Rhodes Roberts

IN FLORIDA, agriculture is the major user of land and water; it must be a key part of the solution to water problems, either through voluntary collaboration and alternative strategies or, if not, through regulation. Agriculture in Florida and other states views new governance systems with optimism tinged with skepticism. In the past too many agricultural producers have been faced with adversarial regulation in water conflicts; they know that this rigid order has not found acceptable solutions for anyone. Some have accused agriculture of being defensive and protecting the status quo. Others say agriculture responds only to significant incentives. I believe that agriculture has significantly shifted towards adaptive strategies through both desire and necessity.

Representation

Gaining true representation is a slow, tedious, yet necessary process. Every landowner and the general public must be involved if lasting solutions are to be crafted. Success hinges on identifying all stakeholders and involving them in deliberations, along with clear communication, incentives, and penalties for lack of progress. For example, in the Suwannee River Partnership, in which I served as chairman for six years, it started with the vision of water management authorities, then three to five years to gain the acceptance of local, state, and federal regulatory authorities, and even longer to achieve complete participation by all industry groups and sometimes skeptical landowners. From the initial informal cooperation in 1994 to a formal memorandum of agreement in 1999, the Suwannee River Partnership was an effective group of 46 federal, state, regional, and local government agencies, as well as industry groups and private associations, that worked together to improve water quality in the Suwannee and Santa Fe River Basins (Roberts 2003b) with plans to incorporate the Georgia portion of the river basin in a multistate operation (Suwannee River Water Management District 2003c).

Another excellent example of representation strengthened by collaboration, public and private partnership, and scientific and technical input, is the beef cattle management practices of the Lake Okeechobee Protection Program in the Everglades system. State government assisted landowners in assessing their operations based on current scientific knowledge and federal agency recommendations. Industry actively proposed Best Management Practices (BMPs) of water management structures, alternative water sources, buffer strips, fencing to reduce water access, and measures to reduce soil erosion and sediment transport. Environmentalists even applauded the endangered species preservation by the program. Use of agricultural lands has also become an important part of hydrologic system maintenance (Missimer 2003). By January 2004 over 200,000 acres of beef cattle production in the four priority basins around Lake Okeechobee had implemented BMPs, with the University of Florida working to provide scientific data. Plans are reevaluated every three years, with constant public learning and changes in practices.

Process Design

How do you design a process to manage the complexities of agriculture? How do you ensure full participation from the smallest to the largest stakeholders, both public and private? The complexity and diversity of agriculture itself presents a challenge. Agriculture is not homogenous, particularly in Florida where over 350 individual crops or animal products are produced (Hodges and Mulkey 2003). Agriculture and forestry use almost 52 percent of Florida's 35 million acres (FDACS 2003), and the individual entities using the land vary dramatically in size and complexity. In 2003, Florida had 44,000 farms on 10.2 million acres ranging from a 1–5 acre niche to operations over 100,000 acres. Some crops are harvested three times a year, while others, such as citrus, are produced for 20 years. Marella (1999) reported USGS estimates that agriculture uses over 48 percent of Florida's total fresh water withdrawals, with public supply estimated at 30 percent. Florida's Agricultural Water Policy (Bronson 2003) estimates that 40 percent of the water pumped for irrigation is subsequently returned to Florida's hydrologic system; the amount depends upon many factors including the type of irrigation system, site-specific conditions, and point of discharge. Estimates of future agriculture water use indicate either static or decreased needs. Micro-irrigation systems increased from 53 percent of 770,000 acres of citrus in 1991 to 80 percent of 830,000 acres in 2001, with an average saving of 140,000 gallons/acre/year (OAWP 2004). Significant improvements have also occurred in the rapidly expanding nursery and ornamental plant industry, as well as in vegetables and other fruits.

Agriculture has learned the hard way that cooperative solutions have a much better chance of success with careful design and full involvement of users. Cooperative, non-traditional solutions need a stamp of legislative or regulatory legitimacy to be accepted as replacements for traditional actions. Long-term solutions are more readily achieved with incentives.

Agriculture has shown a willingness to adapt and to deal with new stresses to the water system, including changing regulatory trends and rapid urbanization.

New institutional bodies have been created to allow agriculture to take on a more proactive and adaptive role. Led by the agriculture industry, in 1995 the Florida Legislature established the Office of Agricultural Water Policy in the Florida Department of Agriculture and Consumer Services to improve communication between agriculture and local, state, and federal agencies on water quantity and quality issues, and to establish water policies involving agriculture. Subsequently all agriculture sectors worked together on a comprehensive agricultural water policy for Florida (Bronson 2003).

Scientific Learning

How do we scientifically measure incremental, continual progress towards a quantitative goal, when we must deal with scientific uncertainties and longstanding fixed dependencies? In agriculture, the carefully crafted solutions for water conflicts are often theoretical, mathematical models; there may be insufficient real world data to verify progress when they are implemented. Accountability of all participants is essential and must be documented. Agriculture is learning the hard way that you must have sufficient scientific data to determine the problem, find the means to solve it, and verify their effect. Measures to resolve water conflicts include significant water conservation as well as reduced use of fertilizer and other chemicals.

Public Learning

From agriculture's perspective, the challenge of public learning is to teach the consequences of production and public practices, communicate the measures required to ensure future use of natural resources, and to instill shared values and goals in urban dweller and agriculture alike. Faced with few alternatives and with deepening concern, agriculture leaders have resorted to legislative actions on water conflicts. For example, after years of discussions and accusations, through the efforts of agricultural and other leaders, the 2000 Florida Legislature enacted the Lake Okeechobee Protection Act (*Fla Stat* Chap. 00-130) committing the State of Florida to restore and protect Lake Okeechobee, and reconciling multi-agency efforts and competing resource users. Government's role has become more intrusive and complicated over the past decades. Large state government land purchases, along with neglected management, has allowed intrusion of invasive species and further complicated any solutions to water conflicts. Government has even turned to agriculture for management solutions.

Problem Responsiveness

Problem responsiveness is perhaps the most difficult of the five adaptive governance challenges for agriculture. Proactive leadership and voluntary consensus can achieve goals in a dramatically shorter time than traditional regulation. Even if that were a consensus premise (which it is not) we would still need consistent interactions, rules, and interpretations to enable agriculture and other parties to work as partners. For instance, citrus requires a 20-year capital investment in land and infrastructure, and owners find it difficult to deal with arbitrary interpretations or unstable situations. Other authors in this volume have stated that cooperative outcomes depend on the willingness and ability of those in power to reward or punish; this is not consistently true. In the Suwannee Partnership, it is true that many cost-sharing incentives were offered to landowners. Yet, in each instance, the individual grower had to pay significant amounts to participate. The "carrot" was there, true, but for many there was no "stick" or "penalty" because the vast majority of willing participants were not covered by any permit requirements or regulatory regime. Over 80 percent of landowners in the watershed basin are becoming participants, while only an estimated 3 percent would have been subject to strict regulatory restrictions under current law. The incentives of peer recognition, true stewardship, and collegial solutions were evident even in the absence of a penalty. Yes, there is the potential for penalties if improvements are not made; however, this is a future threat, not a present danger.

Florida has a plethora of independently functioning, fragmented local, state, and federal agencies, organizations, and citizens involved in water issues. As a former government regulator, I do not believe that existing regulatory structures are fundamentally insufficient, just that individual agencies are historically inadequate to resolve water conflicts quickly and permanently. I am pleased to say that coordinated decisions are now commonplace, and that Florida water governance and agriculture are accomplishing consensus goals, albeit more slowly than most of us would prefer.

Agriculture's Role in Solutions

Agriculture realistically knows that it is not first among public water use priorities. When this is laid against the large percentage of land mass and water volume consumed by agriculture, they have strong incentives, and a unique ability, to foster adaptive governance in water conflict. The heritage of flexibility, independence, and multigenerational stewardship can make agriculture an able partner. People both in and out of agriculture must sort out clashes, trust each other, conduct critical scientific research, and strengthen any new institutions or mechanisms that offer deliberative and adaptive progress. In the end, Florida and agriculture have to keep peace with the independence of the landowner and respect the property rights of the individual, yet realize that the common good of all must be met. Consensus and collaboration are the desired methods. Classic regulation will still serve as the "stick" whenever the "carrot" of new adaptive governance is not achieved.

Part III
Researchers' Perspectives

CHAPTER 14

Resource Planning, Dispute Resolution, and Adaptive Governance

Lawrence Susskind

THE CASE STUDIES prepared for this volume are very revealing. I find, hidden here, five key features of a "system" for allocating and managing water resources in Florida (and elsewhere in the United States). First, above all, in America we depend on government to protect us from ourselves when it comes to deciding whether and how to use water. If we all sought to maximize our own advantage, we would soon be left with nothing to drink. Presumably, because we are not able—as water users of various kinds—to plan ahead or work out our differences face to face or case by case, we have assigned responsibility (and authority) to numerous government agencies to make and enforce decisions.

Second, we seem to put the highest possible priority on maintaining historical patterns of water use. Presumably, the way we allocated water rights and responsibilities in the past is how we ought to allocate them in the future. Indeed, we have a set of property rights that are meant to trump any new interpretation of what might be in our collective best interest.

Third, while science is mentioned quite often, and reference is made to technical inputs of various kinds, politics seems invariably to outrank science as the key factor in water resource management. Scientific uncertainty, which is inevitable given the complexity of the ecosystems involved, is typically the justification for allowing politics to supersede science as a basis for decisionmaking.

Fourth, notwithstanding the first three features of the system, we seem to believe, as Oliver Wendell Holmes once noted, that "general propositions do not decide concrete cases." Thus, we have created an elaborate administrative process within which all the general rules and policies that exist on paper must be constantly interpreted. It is at this level, not at the level of policymaking, that most water disputes arise. And it is at this "local" level that stakeholders and policymakers have more influence than scientists or other experts. The amount of time devoted to reconciling conflicting views in particular cases is substantial and increasing.

Finally, there seems to be blind adherence to the notion that if we just manage our water resources carefully and take advantage of technological innovation as it occurs, we will be able to meet all our future water needs.

As the Florida experience indicates, this system is not working. It does not encourage the comprehensive state and multi-state planning required for effective long-term water conservation. Each piece of the existing institutional apparatus is set up to defend its prerogatives. Thus the status quo triumphs. Moreover, the existing system offers insufficient direction when specific disputes need to be resolved. Disputes are often decided—administratively, legislatively, or in court—in ways that ignore important stakeholders. Finally, the system appears to be doomed in the long run because the emphasis is not on learning how to do better or how to become more sustainable. There are a variety of ways in which the tools and techniques of consensus building (which subsumes traditional dispute resolution approaches like mediation) could be used to bring about a shift from the current system to one based on the principles of adaptive governance. The rest of this paper spells out what this shift would require.

Minimum Conditions for Effective Adaptive Governance

A distinction should be made between adaptive management and adaptive governance. The former is an approach to the conservation and utilization of natural resources (like water); it seeks, through experimentation, to maintain ecologically sustainable levels of use. The latter involves political choices, by which I mean the specification of public policy objectives, the allocation of government revenue, the imposition of regulatory controls, and the allocation of gains and losses necessary to achieve political equilibrium regarding levels of water quantity and quality (in particular places), whether these are ecologically sustainable or not. We can have adaptive management (if decisions are left entirely in the hands of expert agency personnel) without trying to achieve adaptive governance. We cannot, however, have adaptive governance—that is, government decisionmaking regarding use of natural resources that involves all those affected in an effort to find an informed and stable consensus—without directly involving stakeholders (not just their appointed representatives) in compiling and reviewing scientific and technical information as they build consensus on the difficult political choices (trade-offs). Adaptive governance seeks an informed political consensus on how best to allocate resources, a consensus predicated on public learning.

Four minimum conditions must be met for adaptive governance to meet the challenge of problem responsiveness—ecologically sustainable resource management—while satisfying political imperatives. These can be derived from several decades of scholarship and reflection on practice (Lessard 1998; Norton and Steinemann 2001; Wilhere 2002). First, decisions regarding competing uses of water must involve appropriately selected and prepared stakeholders, as well as elected and appointed decisionmakers, in ways that accommodate both short- and long-term interests in a sustainable fashion. If this challenge of representation is not met, it is unlikely that those affected will treat political choices as legitimate. This may mean defining access to clean water (for drinking, agriculture, and recreational purposes) as a constitutionally protected right. Until that happens it will mean, at the very least, guaranteeing all stakeholders that they can select special representatives to participate in resource planning, policymaking, and dispute resolution. While such ad hoc groups can play a key role in building the political consensus, these efforts must only supplement official decisionmaking (by relevant statutory authority). Otherwise accountability and trust will be lost.

Second, those who need assistance to present their views in ad hoc problem-solving forums (not just in formal administrative hearings or courtroom confrontations) must have adequate technical support. This may require water resource planning and management agencies to adopt new forms of joint fact finding (Susskind and Cruikshank 1987). Unless all stakeholders have a hand in selecting the experts involved, framing the analysis, and interacting with the experts after they have produced their analyses, they are not likely to trust whatever assessments are produced. Without a high level of confidence in these assessments, it is not possible to produce credible agreements.

Third, to meet the challenge of process design, all supplementary problem-solving processes ought to be assisted by teams of neutral facilitators selected by the parties (Susskind and Cruikshank 1987). Participants in such collaborative processes must be allowed to question prevailing policies and not be forced to operate within a rigid policy structure. This will require a consensus-oriented approach to formulating water policy and resolving water disputes, as well as a shift away from rule-bound, bureaucratic hearings. The products of such ad hoc policymaking and dispute resolution should be directed to convincing elected officials who are, in the end, the only ones who can be held accountable for public resource allocation decisions. However, any elected body that ignores the consensual proposals of a truly representative stakeholder group that has done its homework does so at substantial political risk.

Finally, defining areas of disagreement, and experimentation and collaborative monitoring to resolve these disagreements, must be the primary responses to scientific uncertainty. Big decisions ought to be put off until all parties have had a chance to study the preliminary results of small-scale experiments aimed at reducing uncertainty to acceptable levels—as suggested in Roeder's chapter on the failure of the legislative consideration of the Aquifer Storage and Recovery issues.

These case studies seem to suggest that there has not been even one water dispute in Florida in which all four of these conditions—involvement of self-selected stakeholder representatives, adequate technical support provided to all parties, neutral facilitation, and explicit experimentation and joint fact-finding—have been met. So the questions that needs to be answered are, "What will it take to accomplish a shift to adaptive governance of Florida's water resources, and what will happen if the state does not move in this direction?"

Problem Responsiveness: Why is Adaptive Governance More Likely to Produce Sustainable Results than Current Practices?

A commitment to adaptive governance will not automatically produce better planning for sustainable use of water resources. Additional commitments are also required. Coordination across a range of specialized activities is essential. Land-use regulation, tax policy, infrastructure (particularly transportation) investment, the setting and enforcement of health standards, budget allocations, and related policy choices must all be properly aligned before water resource planning objectives can be met. Commitments (in the form of specific policy and performance objectives) and associated financial allocations need to be maintained over time. If fluctuations in political leadership lead to constant changes in objectives or reductions in dedicated fiscal resources, it is unlikely that water resource management objectives can be met. Finally, ongoing political support is crucial. Without the continued backing of key constituencies, it is unlikely that initial commitments to achieve sustainable water resource management will be sustainable.

Adaptive governance ensures flexibility. While a commitment to long-term water resource management goals is necessary, flexibility in how those goals are pursued is equally important. Adaptive governance may rely on legislation to set explicit goals and performance standards, but use market mechanisms and other flexible tools to implement them. Without government imposition of goals and performance standards there is no way for tradeable permits or other market-oriented devices to work. Externalities can only be internalized and prices can only be set at the right level if legislation provides a stable context. For stakeholders to hold elected and appointed decisionmakers accountable, the flexibility of adaptive governance must be coupled with a commitment to legislated goals and clearly stated performance standards.

Mandates for stakeholder involvement are vital; otherwise, elected and appointed officials are likely to find too many reasons to minimize meaningful public consultation and joint fact finding. It is easy to argue that there "are too many groups to allow just a few to participate" or "the groups that step forward do not necessarily represent the full range of public concerns" or "the public interest is more than the sum of self-interested stakeholders." These are convenient excuses. Moreover, most elected and appointed officials have a hard time distinguishing between traditional forms of lobbying and the emerging practice of collaborative decisionmaking. They realize that information has to be provided to citizens, so they are willing to hold public hearings or circulate reports. And they are aware that trust in government requires at least the appearance of responsiveness. So, through the use of polls and focus groups they try to demonstrate that they are listening to public concerns. But most are not aware that "best practice" in public consultation now involves self-selection of stakeholder representatives, joint fact finding that brings stakeholders together to interact recurrently with experts that they have helped to select, and collaborative problem solving designed to produce written agreements that detail the ways in which the needs of all stakeholders can be met while resources are managed sustainably. These requirements have to be clear and authoritatively mandated to ensure that they are met.

What is Adaptive About Adaptive Governance?

Learning about complex systems can only occur through purposeful experimentation. Such learning, if structured properly, can inform timely changes in rules (laws, policies and regulations), norms, incentives, organizational behavior, and organizational structure. These, in turn, are necessary to help agencies and organizations achieve their stated goals. Indeed, that is the point of organizational learning as emphasized in adaptive management. Without continuous and effective adjustment, it is unlikely that public agencies will be able to accomplish their objectives in the face of ever-changing circumstances. And without the full-fledged engagement of stakeholder representatives, it is unlikely that the political will to make such adjustments can be sustained. Establishing and applying a rigid set of policies or rules of thumb in all circumstances, regardless of changing conditions, will inevitably produce sub-optimal and less than sustainable results. Flexibility, however, requires ongoing efforts to maintain political legitimacy.

Experimentation, of course, must also be coupled with careful and continuous monitoring for the goals of organizational learning to be served. If there is not an explicit effort to test planned changes in thinking or practice, or to explore the impacts of adjustments against objectives, agreement on what works and what doesn't (and, therefore, what should be done differently in the future) is unlikely. Further, the results of technical assessments need to be interpreted in light of the political interests and values of the stakeholders, along with government agencies and elected officials. Thus, continuous monitoring and interpretation of experimental results must be organized not just as technical tasks, but as political activities (that imply a distribution of gains and losses). So experimentation with the design and operation of water management systems must involve both monitoring and interpretation through a political lens—that is, it must be linked with effective public learning.

In sum, adaptive governance implies continuous adjustment in organizational mandates and routines in response to experimental learning about selected systems, as well as overt consideration of the implications of pursuing one set of objectives over others. In water resource management, adaptation ought to involve experiments aimed at testing new pollution control technologies and resource management arrangements (like pricing techniques), as well as experiments with the allocation of enforcement powers that alter the advantages and disadvantages granted to different geographic and socioeconomic groups.

Why Does Adaptive Governance Produce Better Resolution of Water Disputes?

Adaptive governance produces "better" resolution of water disputes because it yields agreements that are viewed as *fairer* by all the stakeholders; *more efficient*—from the standpoint of independent analysts—because they produce results that meet most of the important interests of all relevant parties while minimizing the investment of time and money required; more *stable* (meaning that they remain in place even as political leadership fluctuates because they constitute meaningful long-term commitments by all key constituencies); and *wiser* because the best possible scientific advice available at the time is taken seriously (Susskind and Cruikshank 1987). Agreements with these characteristics are only possible when the challenges of representation, process design, scientific inquiry, public learning, and policy implementation are met.

Theorists in the dispute resolution field have demonstrated quite clearly how *representation* in deliberative processes associated with adaptive governance ought to be handled (Susskind et al. 1999). Convening agencies need to employ qualified neutrals to undertake formal conflict assessments to determine which categories of stakeholders need to be involved. Stakeholder groups themselves must caucus to select their own representatives with a clear sense of the responsibilities of participation. Representatives need adequate funding to participate effectively and to maintain constant contact with their constituents. Representatives must be held accountable for building support among their members for negotiated agreements that emerge from collaborative dialogues. Hard-to-represent constituencies may need to be represented by surrogates. The entire process of consultation must be given time to produce meaningful results.

The dispute resolution field has also addressed the challenge of *process design*. The key to effective public dialogue is a commitment to consensus building rather than reliance on majority rule. Hundreds of resource management disputes have been resolved by consensus over the past few decades. Ad hoc efforts to engage all relevant stakeholders have produced agreements that elected officials have been pleased to implement. Of course, this does not mean that everyone has gotten everything they wanted. On the contrary, most stakeholders involved in collaborative water resource management efforts have been somewhat disappointed. But they have achieved fundamentally satisfying results that ensure all parties a better outcome than they presume they would have if negotiations had failed. A decision rule requiring consensus encourages all groups to work hard to find agreements that not only satisfy them, but meet the interests of others as well. The role played by professional neutrals (facilitators and mediators) in these efforts is crucial. Consultative processes run entirely by government agencies have been much less effective than those managed by professionally trained neutrals selected by all the stakeholders involved (Susskind et al. 1999).

Consensus processes need ground rules acceptable to all the participants. Consultative groups need to accept the fact that the products of their work must take the form of proposals, not decisions; agencies that encourage collaborative problem solving do not (indeed, they are not legally allowed to) relinquish their decisionmaking authority. On the other hand, if all key groups (including agency staff) have been involved in a highly visible public dialogue that has produced an informed consensus, there is little reason for an elected or appointed body to turn its back on such agreements. Only if some group can demonstrate that it was inadvertently overlooked, or that a collaborative process ignored important scientific evidence, or that some other legal or constitutional constraint was violated, should a convening body ignore a consensus agreement.

The dispute resolution field has spelled out the best ways to incorporate scientific or technical advice into collaborative consultations. Joint fact finding (JFF) requires stakeholders to specify the information they want to collect and the help they want collecting it. Unlike advocacy science (so typical of what happens when water disputes end up in court), JFF assumes that technical advice is sought simultaneously by all the stakeholders (including the convening agencies) acting in concert. While each participant reserves the right to interpret technical results in light of its own interests, it is rare that the product of a JFF process is dismissed entirely. Moreover, JFF puts a premium on expanding the technical capacity of participants through facilitated face-to-face dialogue, with support from technical experts. It avoids the pitfalls of advocacy science, which often tempts all parties to put aside technical advice when it seems that advocates can find experts to say almost anything. Instead of being confronted with scientific reports prepared by advocates, participants in collaborative processes interact with technical advisors whom they themselves helped to select, and receive answers to questions that they raised. Participants are given a chance to understand limitations on the analytic tools employed by their technical advisors. They are briefed on the inherent uncertainties involved in all such analyses, as well as the sensitivity of assumptions that technical advisors are forced to make.

The approach to *public learning* and policy implementation advocated by the dispute resolution field hinges, in large measure, on the idea of contingent agreements. Rather than fight about whose forecast is correct, participants in collaborative processes are encouraged to develop complex agreements that spell out alternative courses of action (and assignments of responsibility) keyed to different futures. For example, the participants in the case studies on water quantity might have agreed that if certain thresholds of water use or water loss (measured in specific ways) are reached at any point in the future, agency policy would need to change in pre-specified ways, shifts in water allocations would automatically occur, and different groups would be required to alter their activities in pre-agreed fashion. Groups confident that such thresholds will not be crossed in the future ought to have no problem agreeing to such contingent conditions. Thus, rather than agree on whose forecasts are correct, parties can agree on a package of contingent commitments.

The results of particular dispute resolution (or adaptive governance) efforts must be folded back into general agency practice in order for problem solving to improve over time. Unfortunately, not many public agencies set aside time to reflect on what they have learned—particularly from their own experience—or how they ought to change their practices in light of such findings.

What Is the Alternative to Adaptive Governance?

The alternative to a consensual, adaptive, stakeholder-driven approach to managing water resources is continued use of traditional bureaucratic administration, legislative oversight, and reliance on the courts to resolve specific disputes. These case studies indicate that most water planning and disputes at the state and local level are handled by elected and appointed officials, with an overlay of "public participation" to ensure accountability, with the courts providing a final tribunal if citizens think an agency has acted incorrectly. Federal and state agencies set standards legislatively, administrative actors work hard to interpret and apply these rules in a fluctuating context, and anyone who claims that he or she is being treated unfairly is invited to seek judicial redress. Ultimately, if citizens are not satisfied, they can elect new officials or lobby for new legislation. Of course, under these rules, there is no guarantee of direct involvement by any particular set of stakeholders, especially those with the most to lose. Nor is there a guarantee that agency policies will change in response to expressions of public concern, scholarly criticism, or polling results. Only a bare majority is required to re-elect each legislative body. The minority is promised nothing but an opportunity to compete in future elections, where their concerns may or may not matter to other voters. And the weight given to scientific arguments or evaluations can vary enormously, depending on the priorities of the officials involved.

Our existing system of representative democracy does not put a premium on getting water resource management decisions right in either the short run or the long run. For example, each water district in Florida can decide how it will meet increasing demand. Whether the overall impact of each of these decisions on the entire state is sustainable, however, is not a question that key stakeholders will necessarily get to ask or help to answer. In the short run, protests such as those documented in some of the case studies do not necessarily change agency policy or practice. Moreover, the courts are increasingly unwilling to contest the substance of administrative decisionmaking, requiring only that procedural safeguards have been met. By the time another election rolls around, environmental damage may be irreversible. In the long term, there is no way to ensure consistent commitments to sustainable water resource management as long as we rely only on traditional administrative decisionmaking and judicial review.

It seems odd that we expect elected officials to be accountable to all of their constituents when only a fraction of eligible citizens bother to vote and only a majority of those are required to elect a candidate. While an unhappy minority is always free to express its opinions, there is no guarantee that these interests will ever be considered. Representative democracy, at least as it is practiced in the United States, assumes that appointed staff with technical background will inform elected officials of the likely impacts of policy choices. These elected officials are expected to weigh this input, along with a wide range of political considerations, and then make wise decisions. It is not clear, though, why elected officials should attach appropriate weight to scientific input, such as the analysis of water resource allocation options produced by engineers, economists, or ecologists, when the political demands of constituents pull in other directions. It is not clear why we are confident that the courts will play a successful custodial role when judges have little or no scientific background, judicial appeals are limited in scope so that only a narrow set of questions may be asked, the high costs of litigation mean that only those with substantial resources will be heard, and precedent trumps all creative efforts to resolve disputes in ingenious ways.

Since the mid-1960s we have supplemented the machinery of representative democracy with public participation requirements like administrative hearings and transparency requirements like the Freedom of Information Act and state "sunshine" laws. These, however, are equally flawed. They do not guarantee that all stakeholders will be heard or that they will have the technical support needed to comprehend the issues. Public participation requirements seem to equate representation with organized interest group involvement when we know, for a fact, that a great many stakeholders are not part of existing advocacy organizations. Moreover, the practice of public participation as it has evolved over the past four decades seems to expect conflicting interests to more or less cancel each other out.

Finally, our system of representative democracy—even with various public participation add-ons—assumes we can count on the media to help citizens educate themselves about the issues being deliberated by their elected representatives. Unfortunately, with water issues in particular, increasing technical complexity and the confusing interaction among decisions at multiple levels of government far exceed the capacity of most news outlets to explain what is going on and what various policy options imply. The very few news organizations that appointed skilled environmental specialists in the 1980s and early 1990s have by now dropped such coverage. The average "story" in the newspapers, or on the radio or TV news, offers little insight into the scientific or technical considerations at stake. Thus the average citizen can remain informed only if public agencies independently maintain sophisticated Web sites, host elaborate public educational campaigns, or publish a steady stream of information in forms that different constituencies can readily access. For economic and other reasons, this rarely occurs.

Conclusions

I have no doubt that almost all the elected and appointed officials responsible for making decisions about the use of water resources in Florida want to serve the public, conserve natural resources, and fulfill their statutory obligations. The system within which they are working, however—including the agency structure defined by Florida law and the general organization of representative democracy—is rigged against them. To ensure sustainable results, a commitment to stakeholder involvement that goes beyond the normal role played by civil society is required, along with a meaningful shift to the practice of adaptive governance.

Until such a shift in attitudes and practices actually occurs—one that puts a premium on the creation of new ad hoc forums that conform to the requirements laid out in this chapter to bring all interested stakeholders together to engage in joint fact finding and consensus building—water resource management planning and typical dispute resolution efforts of the sort used in court are likely to yield disappointing results. Although Florida has been a leader in the use of professional dispute resolution in certain local contexts, this has not yet been extended to the water sector. Moreover, what is required, at a minimum, is state legislation that mandates that all agencies involved in water resource management make a commitment to involve all stakeholders in adaptive governance based on consensus building in all of their activities.

CHAPTER 15

Policy Analysts Can Learn from Mediators

John Forester[1]

IN THE FLORIDA WATER controversies, the disputing parties sought diverse interests. Competing municipalities hoped to use groundwater supplies to support development, to conserve water supplies to avoid shortages, and to improve water quality for human consumption. Employers (and some employees) opted to sacrifice water quality to support industry and protect jobs. Boaters and sportspeople sought water resources for recreational purposes. Environmentalists worked to promote ecological sustainability.

As in other policy disputes, these parties' claims were not only numerous, diverse, and ambiguous, but contradictory. And, they were not only logically contradictory, but expressively and dramatically contradictory too: The parties argued aggressively for their claims in many arenas—the press, political campaigns, courtroom battles, public educational campaigns, and more. But vigorous argument can escalate: it can fuel exaggerations of needs, overly dire warnings, and personal antipathies.

All this poses a problem for students of public policy and citizens alike: when we can expect the parties in public policy disputes to misrepresent their interests—with the best intentions, of course—how can we then evaluate their negotiated policy agreements, or policy proposals in general? What are policy analysts to do when learning how parties misrepresent their interests is anything but simple? What are we to do, for example, when adversarial hearings or an issue-hungry media inflame opposing claims rather than encourage a search for cooperative solutions? If our efforts to create adaptive policy measures are not to be held hostage to such exaggeration and posturing, we need both to explore the many obstacles that keep citizens from learning about each other's interests and to suggest practical institutional remedies: carefully crafted process designs to help us all overcome these obstacles to public learning.

These same challenges face every practicing mediator of public disputes who wishes to work with diverse and conflicting stakeholders and help them satisfy their actual interests and not simply continue longstanding poses of public attack and counter-attack. In complex policy disputes, likewise, we need to explore parties' interests carefully instead of blinding ourselves by simplistically and pre-emptively focusing only on each party's loudest demands. This much is easy to say, but what does it mean for public policy and planning analysis?

Policy analysts, and mediators too, must listen in contentious policy settings for far more than meets their ears; stakeholders have good reasons at times to be less than forthcoming about their interests (Forester 2004b). As they do the best they can for their constituents, they may try not to be overly revealing. And they may be less than wholly candid as they present themselves to the press—or even to allies who seem to be in danger of relaxing their support.

So in policy controversies and political wrangling like the Florida water policy disputes, parties can reasonably worry about focusing narrowly on key threats, about countering the unwarranted claims of their opponents, and about trying to build or maintain public understanding and support. But notice that exploring real possibilities of collaboration does not seem high on anyone's list: in a competitive liberal culture and a complex jurisdictional world of conflicts among agencies, municipalities, and varying levels of government, these many parties seem to presume that adversarial and competitive strategies make up the only game in town—and who can blame them?

In what follows, I explore citizens' difficulties in learning about one another's interests and in devising adaptive policy measures in response. Along the way, I explore the ways that policy analysts can learn to design better adaptive governance processes from public dispute mediators, practitioners who routinely work with conflicting parties who may misrepresent what they are really seeking.

Presumptions Hold Us Hostage

As readers of any case studies we try to imagine what the key actors in those cases have been thinking, feeling, seeking, hoping to defend and achieve. But that problem of imagination doesn't stop, or even start, with readers, for the actual parties themselves very often have limited, indirect information about one another. "Engineers only care about the numbers, and not the people problems," one will say, and another distrusts state agencies and knows ahead of time what they really think.

When mediators of public disputes describe their work, they stress the presumptions that parties bring about each other—what they really care about and what they will and won't do. Gordon Sloan, a Canadian lawyer turned mediator, described his initial conversations with parties and their pre-mediation attitudes about one another:

> They're each saying *exactly* the same thing about the other. That's a piece of information that they should know. It's handy ... to tell them that, when they say to you, "You'll be able to trust what we say, but there's no way you can trust anything they say."

> It's great to be able to say to them in response, "You won't believe this, but they used *exactly* the same words to describe their view of you."
>
> They're amazed. "They did?!? *They* don't think *we're* accountable?"
>
> They discover that there are all kinds of assumptions that one value system makes about the other that have to be debunked. (Forester and Weiser 1996, *16*)

Sloan's remarks suggest that indirect information about other parties does not serve anyone very well. Each party, he reports, can be quite confident about the other's intransigence and lack of interest in serious, collaborative talks that might serve their interests. But, he tells us, they are often wrong: they are too quick to judge, presuming each other's indifference or antagonism, and so the net result can easily be—without a mediator's skillful intervention—that the parties' presumptions keep all of them needlessly ignorant, systematically lacking the organizational and institutional intelligence they need to resolve their disputes adaptively. These presumptions pose abiding problems not only of public learning, but of institutional and political process design as well.

Another mediator of public disputes, Wallace Warfield, suggests how an intermediary can help:

> It's interesting when we do this, because we oftentimes find, even though these people may have been in dialogue over a long period of time, over a number of years about issues, that in terms of how they hear each other's voices ..., the way they talk to each other is in rather limited, narrow channels. They talk in the language of contention. They talk in ways that they really don't hear what the other is saying. *What the intermediary does is create a new way for people to hear each other.* (Forester and Weiser 1996, *172*) [emphasis added]

Speaking of behavior that he has found common to public disputes, Warfield also tells us of predictable political failures of public intelligence, our institutional (in)capacities to learn about other parties and what they really care about. He suggests that processes like mediated negotiations or facilitated policy dialogues, processes that bring parties together with skilled intermediaries, can promote public learning by creating "a new way for people to hear each other," even after these parties have spent years in which "they really don't hear what the other is saying" because they have been so exclusively talking "in the language of contention."

Warfield and Sloan both warn us that in longstanding public disputes—*Apalachicola* and *Tampa Bay*, for example—parties can easily acquire mistaken and indeed self-defeating confidence about one another's real intentions and interests. No party to the water policy disputes, after all, has direct experience of all the other stakeholders or complete information about them. In contentious situations, when the parties themselves can easily be disserved by secondhand or press-enhanced reports, skilled intermediaries might help those parties learn from and about one another in ways that their own public posturing might otherwise prevent.

Challenges of Representation: Just What Interests are Represented?

The Florida water policy disputes involve technical issues, of course, but they also involve contentious, politicized, and passionate relationships between those who support or oppose further growth, who wish to use or to protect environmental resources, and who try to defend neighboring but competing political jurisdictions. These parties not only feel strongly about their own issues, but they can feel threatened by the intentions of others. Do we preserve groundwater quality by not replenishing our aquifers or do we inject treated effluents into those aquifers? Do we trust the reports that assure us of public safety, or do we worry that the advocates of the injection strategies paid for the reports? Do conflicts of interest jeopardize public learning?

In contentious policy situations, we can hardly expect parties always to be "reasonable," considerate, and even-tempered. So the parties can expect each other to posture, exaggerate, overly stress one point over another, and so on. That may be understandable, but it does not help anyone—either policy analysts or the parties themselves—to learn about the actual interests involved, the actual interests deserving representation and consideration in public policymaking processes.

So if we wish to assess public disputes, negotiations or policy deliberations, we must not only expect and recognize public posturing, opening demands, wishful thinking, positions staked out, and so on, but we must not be held hostage to that bluffing. We must also explore other less obvious interests, "underlying" or additional concerns that are also at stake and might play a role to motivate public action. We need to distinguish predictable posturing from less exaggerated concerns so that we can better gauge actual priorities and truly pressing concerns. Unless we students of adaptive governance can actually address those real interests (including natural resource protection), we can hardly succeed.

We cannot simply assume that parties to policy disputes are perfectly rational, all-knowing, somehow not caught in the trap of being mutually deceptive: the trap of mutually unsatisfactory compromises—if not quite "lose-lose" agreements—as a result of mutual posturing (Lax and Sebenius 1987; Raiffa 1985; Ostrom 1990). Public policy formulation, no less than public learning, thus depends upon an analysis of the real, roughly "ranked" interests that reach far beyond any first few loudly espoused interests (of industry actors, recreational fisherman, or environmentalists, for example), for only then can we begin to assess not rhetorical game-playing but negotiation efficiency, well crafted attempts at adaptive governance, and the actual costs of failing to reach outcomes of mutual gain.

But policy analysts can learn from mediators' experiences with conflicting parties. Consider how practicing mediators might interview stakeholders in policy controversies like the Florida water disputes. Mediators typically need to ask each party about their best alternatives to potentially negotiated agreements and the costs of each, including, of course, their no-agreement options and the associated costs of doing nothing or remaining in the status quo. Assessing alternatives in this way provides a practical picture of power and interdependence: a picture of who needs to "give" in order to "get" something else, and who, in contrast, can more easily afford to keep things the same.

But mediators know that in order to serve the parties to a dispute, they must learn about the parties' interests, even as those interests shift over time. Reflecting on their experience, mediators have observed that many times, strangely enough, the passionately involved parties may not yet have clear ideas about the actual interests of other parties or, surprisingly, about their own. Not only do mediators find that parties often posture and mislead others—other stakeholders and perhaps policy analysts too—but they also suggest that the parties themselves can be internally conflicted or unsure about their own interests (Stiftel 2000; Forester 1999). But how is that possible? Mediators warn us, in effect, that disputes often turn out to be hardly what they first seemed (to anyone) to be about, so that if we listen too literally or too gullibly, to what the parties say they want, that may not serve anyone very well.

As ordinary listeners, we know this, of course. When emotions run high, when trust is low, when pressure is on, when one party fears another, when the relationship is historically stressful: in all these kinds of circumstances we may posture, simplify, exaggerate, minimize, conceal, dig in, dismiss, and in so doing we can easily fail to say clearly—and fail to let others know clearly—what we really care about.

If that much is true for any one of the parties, what happens when they meet? In situations of conflict and mistrust, of inequality and suspicion, each and every party is likely to "come to the negotiating table" with just such practiced wariness, caution, reticence, or even perhaps a deliberate show of resolve, commitment, or muscle. Take one part history of mistrust, add two parts suspicion and display of resolve, and we have a recipe for earnest, if not righteous, opening statements and positional bargaining that hardly expresses what those same parties really care about or hope to achieve (Lax and Sebenius 1987; Susskind et al. 1999, Chapter 12).

Or consider the difficulties faced by representatives of internally complex parties. These negotiators, Bruce Stiftel (2001) suggests, face daunting problems of working effectively at "two tables," one that we can call "at home," at which they try to resolve the ambiguous and internally contested formulation of their own interests, and another "in public," at which they as representatives seek to articulate, defend, and promote their interests in the face of potential adversaries. So a state government spokeswoman may be painfully aware that what she can advocate in a given negotiation depends on a fluid political environment as legislative debates proceed about economic development or environmental protection measures. The Chamber of Commerce may have an old guard wrestling with a younger cohort of emerging leaders, and they may hardly see eye to eye about the real interests of the Chamber. When the stakeholders in a public policy negotiation themselves represent complex coalitions or combinations of other parties, knowing clearly what their own interests are can become quite a challenge for all concerned.

Process Design That Enables Inquiry: Conflict Assessment Questions Mediators Might Ask

So as mediators design dispute resolution processes, they know that they need to allow time and space to listen not just to the parties' initial claims made in opening statements, but also for quite a bit more. They know that what parties say most emphatically can serve many purposes well beyond self-disclosure: the parties may try to protect themselves, to look good, to make peace at home, to focus on past wrongs, to play to public or politicians' sympathies, to extract concessions from others, to counter, intimidate, or dismiss others, to exaggerate needs, to minimize their willingness to give, and much more (Forester 2005). So how can sensitive process design help with these problems?

Mediator Jon Townsend reflects on these difficulties and offers us a provocative observation drawn from his years of practice in the United States and abroad:

> A mediator needs to think like a negotiator because that's what the parties are. The parties are negotiating, or their negotiation-communications have broken down. But they are negotiators nonetheless.
>
> I mean, they may be poor negotiators, i.e. poor communicators: they may not know their best interests. They may not know what their interests are. Most people don't. Most people are positionally-based, right? Be it in formal negotiations, or, if you go to mediation, you usually take a position if you're a party. You usually don't think about what your interests are. (Forester and Weiser 1996, *109*)

Townsend's suggestion that most people in negotiations have not thought clearly about their interests would be less stunning if other mediators did not echo and even amplify his point. Gordon Sloan observes:

> [In multi-stakeholder settings, as they prepare for negotiations, parties] begin to crystallize their own interests. They don't know that yet, because they don't know what "interests" are, but they begin to retreat from positions that they take about the land base, and they *begin to identify specific areas of need, desire, concern, fear, aspiration, expectation*—what I would call "interests," that they've got to be very clear on by the time the negotiation gets rolling. (Forester and Weiser 1996, *17*; Forester 2005) [emphasis added]

Mediators Townsend and Sloan both suggest that in the pressure and heat of adversarial negotiations, parties may often set out "positions" and fail to assess clearly and carefully their own interests that may underlie those positions. When opponents in *Tampa Bay*, for example, press their demands and ostensible needs for water consumption or conservation, thinking positionally seems to save time and focus attention. In stating a position, thirsty users don't have to say when or under what conditions they might make concessions—they can say directly what they want. In setting out that position, they seem not to worry in that moment about what environmentalists want today and what more they might want tomorrow, and they don't seem to worry about the environmentalists moderating their stances or their (apparently unlikely) cooperation. So farmers, for example, take the position that they want to preserve these acres, or a municipality announces its position that it wants to build this reservoir, or the boating lobby fights for the position that it should be able to fish in this river. As these diverse parties state positions, crystallized pictures of what they want, these pictures leave unstated *why* they really want whatever they've said they want. And even if a substantial literature suggests that such strategic behavior can be self-defeating, it has been very hard to resist (Raiffa 1985; Susskind et al. 1999).

But such positional thinking leads not only to simplifying and often oversimplifying demands in the face of political complexity, it also appears enticingly to solve another problem. When cooperation and collaborative communication have broken down, staking out a position boldly claims turf in the face of opposing claims to the same acreage, reservoir site, or use of the river. The more adversarial our interaction, the more I may feel compelled to state my position clearly. The worse our communication, and the less I trust you to listen, the less I will trust you to care about my interests, *why* I want what I want—and the more likely I will be to set out the basics of *what* I want—even at the risk that in doing so I might bury or obscure or even fail to understand carefully those multiple interests I'd really like to satisfy for my own good (Fisher and Ury 1983).

Notice how Sloan distinguished "areas of need, desire, concern, fear, aspiration, expectation" to indicate the complex varieties of interests that parties can have and the ways those interests can be felt. Fears differ substantially from aspirations, for example, and needs and expectations differ from one another—but all these can motivate self-interested action. All of these "kinds of interests" drive parties not just to act, but also to posture if they feel that exaggeration will serve them well. So just as Sloan shows us what he as a practicing mediator hopes to *learn* about each party, he also shows us that "positional" behavior and posturing can hide much from everyone's understanding—both the other parties' and his own.

So if the parties themselves in the Florida water disputes—urbanites and agriculturalists, developers and environmentalists—have had several reasons not to analyze their own interests, we can appreciate that policy analysts, practicing mediators, and analysts of negotiations must all work hard to probe behind espoused positions to assess these parties' actual interests: the range of diverse interests that these parties may have buried beneath their publicly proclaimed and loudly defended strategic positions. Sloan sets an agenda for policy research by differentiating these questions of "need, desire, concern, fear, aspiration, expectation."

Challenges of Public Learning

But consider further why "positional bargaining" can be so seductive, despite its oversimplifications. Why would any party—environmentalist, farmer, factory owner, elected politician, boater—persist not only in hiding their own interests but in failing to learn about the actual interests of other parties, when that gamesmanship threatens to lead to mutually disappointing, "lose-lose" compromises that forego the joint gains that they otherwise might achieve from careful negotiations (Raiffa 1985; Susskind et al. 1999, *467–472*).

We find a clue in the work that mediators often do to build relationships between the parties even before those stakeholders begin to negotiate their differences. Fisher and Ury (1983) long ago urged negotiators to distinguish "the people from the problem" and to learn about both: to learn as much about the people and the relationships involved in a dispute as about its apparent substance. As Warfield put it, mediators try to design processes, to set up transparent interactions, to "create a new way for people to hear each other."

Why might mediators bother to do this? Just *how* they try to build relationships before trying to negotiate or allocate resources matters less for our purposes here than their signal to us *that* they do need to do this. They suggest that how the parties reveal or explore their interests will depend upon their relationships (how adversarial? how cooperative?) with the other parties, and their expectations of those other parties' consideration, respect, listening, flexibility, suspicion, and more.

If environmental groups and industry representatives have decades-long histories of litigation and very short histories of collaboration, they may never ask questions that would simply seem "pie in the sky" to them: "What could we really do if we really worked together? Could we make a case for new public investments to explore technologies that would serve both industry and reduce environmental impacts? What environmental protection initiatives might become possible if a broader coalition of supporters advocated for them?" Each case will be different, of course, but as long as parties *presume* that fighting each other in court is the only way to move ahead, they will forego asking other questions about interests that they might achieve together.

Here press coverage can heighten parties' perceptions of one another as ideological, intransigent, untrustworthy, and worse. What sells newspapers may not be what encourages the good working relationships required for collaborative inquiry and public learning.

Histories of distrust, interagency and inter-jurisdictional competition, and an always changing political and economic picture can all make collaboration difficult, and all these can promote fears that other parties are just "out for themselves," that other parties pose threats to scarce resources. Add press coverage that may not always quote accurately and that might inflame suspicions, and all parties will find it more difficult to talk safely with one another and to think creatively about innovative policy measures.

Without deliberate process design to counteract these problems, we have institutions not for public discovery and learning but for exaggeration and mutual deception. Yet as Sloan and Warfield have suggested, parties can learn about one another and see new possibilities to explore, but only if they are first able to build new working relationships in which they can appreciate the range of interests that each party hopes to satisfy (see Susskind et al. 1999, Chapters 8 and 12).

For one mediator, for example, this meant convening a multi-stakeholder mediation—to make a policy recommendation to a state legislature—as a "study group" rather than as a "mediation," simply to stage the parties' working relationships from the very beginning in a less adversarial way: to pose the challenge to the group not as deal-making, negotiating as adversaries, but as meeting together to learn so they might then act (Forester 2004a). In this case, asking, "What information do we need to collect to understand our options?" replaced "What do you want and why?"

This strategy helped the nonetheless competing stakeholders to listen to each other and each other's rationales for "why we need this information," and it enabled those parties to inquire together, to frame questions and assess data, and to learn together so that they could make the legislative recommendation they were convened to make. Once again, how parties reveal, explore, and address their diverse interests will depend on the forms of stakeholder interaction and the process design that frames those (more or less adversarial) problem-solving interactions.

Parties can learn about one another's interests in part, then, as a function of the relationships they think possible with others. If we think we're at war, different interests will seem relevant to us than if we think we're facing a golden age of cooperation.

Lawrence Susskind notes these contingencies too: parties will consider their interests differently over time as their mediated or facilitated conversations evolve:

> People start this process with needs, desires, wants, concerns, ideology, uncertainty, and interests. I expect people to change—to alter their sense of what they would or wouldn't like to have happen by listening to what other people say.
>
> Learning and inventing goes on, reconsideration goes on, and argument matters... People discover something about their own interests along the way. (Forester 1994)

Here again, we see a scholar-practitioner's suggestion that parties' own sense of their interests may not be given, fixed and clear, once and for all at any one point in a public dispute: public policy disputants can learn as they listen to each other, as they consider new data, as they find new scientific studies, as they learn about strategies that other entities have tried. In any public dispute, then, we should expect the parties' initial announcements of "their interests" to reflect a work in progress— their first word, perhaps, but certainly not their last word.

Notice that another "contingency" involves the presence (or absence) of possible sponsors of facilitated deliberations. The Florida water policy cases beg questions of such assisted deliberations, for surely the interests revealed in adversarial administrative proceedings or in raucous public hearings will be much more partial and exaggerated—and difficult to work with—than the broader and more reflectively considered interests of parties who might participate together in carefully facilitated processes of collaborative problem solving. In facilitated processes, the parties themselves can not just learn more about each other's interests, but they can craft constructive, practical proposals to meet each other's interests—the more they can take advantage of a less immediately adversarial, litigious, and formal "deliberative infrastructure," the more they can meet one another less defensively and so explore together—somewhat protected from the glare of the press and from each other's posturing—proposals for workable policy measures (Forester 2004b).

So if parties to policy disputes have only met one another in settings that reward exaggeration and mutual defensiveness, or if, as often happens, they have rarely really been in the same room together to discuss their concerns, we can hardly blame them for failing to appreciate the range of interests that really matter to each other. Not only will adversarial settings encourage exaggeration and posturing, wariness and defensiveness, they will also make it quite unlikely that the parties will consider their mutual interest in cooperation, actually finding ways to reach mutual gains instead of reciprocally generating lose-lose outcomes, what we might call, "mutually lousy compromises."

What could appear more irrelevant than the hypothetical benefits of cooperation when the parties are investing deeply in litigation, public posturing, and competitive behavior to satisfy what they espouse as their primary interests? But worse than irrelevance follows: when failure to understand each others' secondary interests increases everyone's chances of proposing strategies and policy measures that poorly serve anyone's actual interests, it also becomes more difficult for any party to realize how poorly interest-responsive a given agreement may be, *because they simply don't know what interests are actually at stake.* Not only does the deceptively rational posturing of all stakeholders hamper their understanding of interests and their ability to craft adaptive policy agreements, but the same mutual deception hampers everyone's ability to evaluate policy proposals as well. How can we as observers know how adaptive a policy proposal really is if we do not know the full range of interests that proposal was expected to satisfy? We cannot.

Process Design and Sponsorship: Building Mediated and Deliberative Infrastructures

Predictable problems of gamesmanship are a good reason to explore the use of third party intervention. In California, for example, the Center for Collaborative Policy takes on challenges like that faced by the Florida Conflict Resolution Consortium, influential in *Tampa Bay*, *East Central Florida*, and *Everglades*.

Under the leadership of Susan Sherry, California's Center for Collaborative Policy has established an impressive record of water-policy related work, most famously in the 2000 Water Forum Agreement. That agreement was "a negotiated settlement of interlocking components designed to achieve the co-equal objectives of providing a reliable water supply for the Sacramento region's planned development to 2030 and preserving the fishery, wildlife, recreational and aesthetic values of the Lower American River" (McClurg 2002).

Sue McClurg quotes Susan Sherry: "The water purveyors wanted increased surface water diversions and new facilities. In exchange, the environmentalists got dry-year alternatives, Lower American River flows and habitat projects, water conservation and groundwater management, and assurances that all of these things would happen" (McClurg 2002, 6).

That Water Forum Agreement may well have been an extraordinary accomplishment that took many years of work. But surely the Florida adaptive management cases also show that extraordinarily complex policy disputes call for similarly ambitious, innovative, adaptive, and collaborative policy processes.

This suggests once again that process design, the deliberative infrastructure of governance itself, shapes stakeholders' abilities to learn about their own and other parties' extensive and potential interests. What one party will consider to be "in their interests" will depend in part on the availability of skilled professional third parties who convene and manage safe, well informed processes of collaborative inquiry and mediated negotiations. Lacking such well designed policy-crafting processes, parties will reasonably retreat to defensive and narrow conceptions of their concerns, public learning will be pitiful, and all parties may lose as a result.

Both our discussion here and practical experience like that in California suggest a role for university-based research programs to provide analytical capacity and deliberative settings in which stakeholders can learn about each other, the issues at hand, the costs of mutually inadequate agreements, and the creative options yet to be explored. Providing such analytical and practical capacity could be a significant public service, but of course it challenges the traditional "technical assistance" understanding of research with a still more applied view of participatory research (Greenwood and Levin 1999). Here university extension services could help to provide the infrastructure for adaptive governance conversations that could in turn promote successful public deliberations. The penny-wise, pound-foolish gamesmanship that we see not only in the Florida water wars, but in public disputes in general, suggests that greater public investment in dispute-resolution processes tailored to complex policy disputes could repay their initial costs very handsomely, avoiding costs to industry and the environment, taxpayers and public trust alike.

Assessing Predictable Pathologies of Misrepresenting Interests

We can bring together many of these arguments in Table 15-1, presenting the obstacles to stakeholders' learning about their own and others' actual interests in contentious and complex policy cases. We can expect stakeholders to have interests both in the substance of a policy dispute and in the relationships between the parties to that dispute. They will be interested not only in being able to discharge effluents into the Fenholloway River or in maintaining a given level of water quality of the river and wetlands, but in maintaining working relationships with permitting authorities, legal counsel, and each other as well (making possible the Fenholloway River Evaluation Initiative).

But stakeholders can frame each of those areas of interests—the relationships to be managed and the substantive resources at stake—in either of two ways, or both: first, a party can frame its interest in strategic ways—as a bargaining attempt to get what it wants. Second, that party might also frame its interest in a way that depends on its understanding at a given time of what the available technology allows or what relationships it might ever have with another party.

Table 15-1. *What Obstructs Learning About Other Parties' Actual Interests?*

Parties selectively reveal their cares and interests	*Parties can have interests in* *Relationship*	*Substance*
deliberately & strategically withhold interests as they	• appear accomodating • appear intimidating • feign cooperation	• exaggerate the new • hide priorities • withhold information
contingently (as "context") changes because they	• expect sanctions • expect disrespect • expect little cooperation	• expect technical change • expect environmental change • expect changing resource needs

Note: Following Sloan, "interests" is shorthand for concerns, desires, hopes, fears, worries, preferences, needs—all of which can matter practically, can be significant, and can be more or less deliberately hidden or not even perceived clearly.

If we array these two areas of interests against these two ways of framing and presenting them in any actual dispute, we find the possibilities mapped in the table. Here in the four quadrants, we see four possibilities of presenting—and hearing—any party's interests, and we have four areas in which *any*, or *every*, party can have real difficulty understanding what actual interests another party may have.

To make matters worse, these four areas of difficulty in understanding parties' interests also present difficulties for public policy observers and analysts. Of course, these ways to selectively frame a party's interests make it difficult to evaluate proposed policy measures, because surely one leading criterion of policy evaluation (along with concerns of justice and fairness) must involve how a policy measure does or does not actually satisfy the real interests of affected stakeholders. But if we as parties, or as the public, only partially know the real interests at stake, for any of the reasons mapped in Table 15-1, then we are that much less able to evaluate policy proposals.

Consider each of the cells in turn. In the upper right corner, parties frame their interests in the substance of the negotiation strategically, and so here we find the most conventional bargaining behavior. Believing that they must compete for limited resources in a zero-sum world—so that any increased ability of industry to discharge effluent seems to promise diminished environmental quality, for example—each party may stress their priorities and minimize other concerns, hoping to protect what they care about the most. Although well-known pathologies of bargaining suggest that when everybody postures, everybody risks doing less well than they otherwise might, exaggerating and misrepresenting one's interests are all too common (Lax and Sebenius 1987; Stone 1997).

In the upper left corner, we find different problems. Unable to assume trusting and cooperative behavior by others, parties may frame their interests in the negotiation's relationships strategically, and so here we find parties saying that they wish to join forces, or to be cooperative, or to go it alone, or trying to look confident or resolute or threatening—none of which may be true. Parties can also have strategic reasons not to disclose their interests in desired relationships because they can fear "looking weak" or being left in the cold without allies. Misrepresenting "relationship interests," apprehensive, distrusting and fearful parties may never discuss—may pre-empt discussion of—their joint interests with other parties who seem intent only on meeting them in court (see Argyris and Schon 1978).

In the lower left corner, we find that the ways that parties reveal or discuss their interests depend on the changes they find in significant working relationships. The hope of working together with a regulatory team may go up in smoke as the new team leader seems unable to listen to anyone but himself; or a new state representative kindles fresh hope of cooperation as we see that others share our fears of economic or environmental devastation. Threatening or cooperative, respectful or dismissive, relationships can change with new personnel, new elections, new development initiatives, or changes in the leadership of community or environmental organizations. Parties may be more or less willing to reveal what they really care about and what their priorities really are, and what secondary interests really matter to them.

Finally, in the lower right corner we see yet another set of obstacles to learning about each others' interests, for how we think about and reveal our interests depends in part upon our knowledge of how the world changes around us. New medical or toxicological testing technologies allow us to be more precise about what we feel we need to regulate, or more confident about the safety or the risks of injecting treated waters deep into the ground. As technologies change, as the policy and tax and regulatory environments change, parties may reconsider their opportunities and priorities, and articulate them accordingly in practical policy negotiations.

Conclusion

Table 15-1 maps several ways that parties to public disputes can have good reasons to suspect one another's espousals of their interests. It also summarizes the reasons that policy observers may be skeptical of parties' espousals of their interests. But knowing that another poker player is bluffing hardly tells us what hand to play.

Similarly, because citizens' real interests provide a significant indicator of what we expect good public policy to satisfy, Table 15-1 suggests an agenda for research if we are to understand policy negotiations better in the future. Studying policy conflicts, we need to listen more astutely than ever, well beyond parties' opening gambits, well beyond parties' public espousals. We need to focus less on the interests that parties emphasize and may exaggerate, and more on parties' further concerns, aspirations, needs, fears, and, in a word, priorities. Of course, we need to consider those concerns and interests as politically and even scientifically contingent; not only political administrations but also scientific findings change over time.

The table also provides us with reason upon reason to take process design seriously to enable public learning: to establish carefully facilitated and mediated processes, real functioning deliberative infrastructures of adaptive governance. We know that in contentious environmental and public policy disputes, parties have good reasons to misrepresent, or even fail to look closely at, their own actual interests. They have good reasons to distrust others and to exaggerate, withhold, and treat as not discussable topics of real concern, but those understandable good reasons create public policy pathologies of lose-lose, sub-optimal agreements replete with battered relationships, economic waste and inefficiency, collective ignorance, and environmental destruction.

Those pathologies should move governments at all levels to build carefully designed, well-informed, publicly accessible deliberative infrastructures: capacities enabling skilled analysts to assess conflicts in order to recommend if and when trained mediators or facilitators should convene joint deliberations; capacities to make available non-partisan scientific resources, during and between deliberative sessions, to supply needed information; capacities making available mediators, facilitators, or managers to create safe spaces for collaborative inquiry between parties who will not be held hostage to their own defensive, strategic inclinations. In these new deliberative spaces, parties can learn together, through face-to-face dialogues and joint fact finding, for example, that they can avoid the pathologies of mutually misrepresenting interests, mutually undermining policy innovations, and instead work to make adaptive governance a practically negotiated reality.

Notes

1. Thanks for comments on related materials to David Booher, Judith Innes, Laura Kaplan, Connie Ozawa, Tore Sager, Larry Susskind, and Michael Wheeler.

CHAPTER 16

Leadership and Public Learning

Robert M. Jones

REACHING CONSENSUS on water resource projects may require a new style of leadership. As these case studies illustrate, only leaders who promote public learning can overcome distrust, uncertainty, and doubt. They must deal with representation issues, process design, and how to connect scientific and public learning with adaptive management by specialized agencies. These water resource problems are fraught with scientific and political uncertainty that demands an interdisciplinary, adaptive approach to leadership and problem solving (Lee 1993). Such an approach may bridge the gaps between scientific learning, public learning, management, and policymaking to produce feasible solutions.

This paper explores the role of leadership in these water resource initiatives and in meeting the challenges facing adaptive governance, in particular scientific and public learning. It concludes that applying adaptive strategies within the traditional institutional and political structures that govern water resources will require a creative rethinking of leadership and public learning.

The case studies were selected because they presented instances where no single administrative authority could resolve the issue or manage the resource. They suggest that the traditional leadership of legal, administrative, and political institutions needs to cultivate the skills to convene those with a stake in the water resources decisions and develop consensus solutions.

"Wicked" Water Conflicts and Leadership Challenges

Whiskey is for drinking and water is for fighting.
Mark Twain, 1897

Fights surrounding the water resources in Florida and in other parts of the country have been making headlines for over a century. Rittel and Weber (1973) suggested that "wicked" planning problems often have uncertain boundaries, defy absolute solutions, and are themselves symptoms of larger problems. Adler's (2000) restatement shows the parallels with the contemporary water conundrums described here:

- "Wicked" problems are diabolically complicated and filled with uncertainty.
- Tentative solutions for water resource issues, based on best available but imperfect knowledge, need to be tested experimentally throughout implementation.
- The emotion, politics, and intensity that characterize these conflicts require exhausting lengths of time to build trust and public recognition of the problem, let alone build consensus for collective action.
- Solutions proposed by parties who come and go (or change their minds, or fail to communicate, or otherwise change the rules by which the problem is being addressed) create additional instability.
- Leaders of efforts to find solutions for "wicked" problems often must adapt to emerging political implications that are revealed only after stakeholder consensus and investment of political capital bring forth recommended solutions.
- Assurances to leaders and stakeholders that proposed solutions are both the best and the cheapest will be difficult to deliver, especially if adaptive management is central to the implementation process.

One leadership scholar, Ronald Heifetz, has been exploring the kinds of demands and challenges that complex public problems present for today's leaders and their communities (Heifetz 1994). Table 16-1 summarizes his characterization of types of water resource problems, distinguishing the technical problem (clear problem with a known fix) from the adaptive challenge or "wicked problem" (uncertain nature of the problem and disputes over the right fix). He lists some of the demands on leadership for these different public learning challenges. Heifetz argues that an adaptive or "wicked" problem requires a process of public learning and consensus building on the nature of the problem as well as the best strategies for solving it. This may require individuals and institutions to undergo a transformation in which facilitative leadership is a critical ingredient (Chrislip and Larson 1994).

When faced with the adaptive challenges presented by water conflicts, leaders must promote that transformation.

Collaborative Leadership on Water Resource Challenges: Principles and Tasks

The Florida water resource case studies illustrate various stages and challenges involved in the political and legal transition from reliance on confrontation to collaboration and from first order to second order governance. They highlight both the use and misuse of science and technical information to inform public

Table 16-1. *Technical and Adaptive Challenges*

Problem types	*Nature of problem*	*Example*	*Leadership implications*
Type I Technical Convergent Problem	Agreement on definition; agreement on the fixes	How do we retrofit old water systems for conservation?	• Tend not to require much consideration of values and beliefs. • Resources and expertise usually solve them. • Do not usually require high levels of participation.
Type II Value/Divergent	Agreement on definition; No agreement on fixes	When we have tapped out our currently available groundwater, how do we expand our water supply?	• Require much consideration of opinions, beliefs, values and convictions. • Resources and expertise alone will usually not solve them. • Requires high levels of buy-in by those who have the problem. • The solutions to adaptive/divergent problems require more value- based diagnostics and typically lead to larger shifts and transformations.
Type III Wicked/Adaptable	No agreement on problem definition and no agreement on the fixes	Who should have "first call" on the cheapest water available now and in the future?	• There is broad disagreement on what "the problem" is, competing solutions that activate a great deal of discord among stakeholders, and the power to define both problems and solutions is diffuse or contested. • They are driven by conflicting values but they have deep, long, and often nasty histories that are "remembered." • Communication is virtually cut off. People communicate at a distance and through extreme positions.

Source: Drawn from Heifetz (1994).

learning in navigating this transition. These challenges point to the critical importance of leadership strategies that can create a shared public vision of change and joint strategies that address concerns beyond any particular water resource stakeholder's purview. This is seldom an easy task. Finally, they make clear that institutions need more leaders who can facilitate consensus building.

The case studies suggest that effective leadership for adaptive change often demands a blend of traditional and new collaborative leadership skills. For example, in a more traditional vein, bipartisan legislative support made possible the *Everglades* Restoration. That legislative cooperation and coordination culminated in the nearly unanimous passage in 2000 of implementing legislation at the state and congressional levels, which provided authorization consistent with the stakeholder consensus plan. The result was a commitment of $8 billion to the 40-year restoration project. Smart political leadership was bolstered by a consensus solution shaped by the key affected interests and widespread popular support in Florida and the nation.

In their work on collaborative leadership, Chrislip and Larson (1994, *138–141*) studied over 50 community leadership case studies and distilled the following collaborative leadership principles:

- Collaborative leaders inspire political and personal commitment and action.
- Collaborative leaders function as peer problem solvers.
- Collaborative leaders build broad-based involvement in the collaborative enterprise.
- Collaborative leaders work to sustain hope and encourage participation in the consensus building process.

The case studies suggest that there are key leadership tasks in the initiating and convening stages, in the consensus-building process, and in the implementation of the results and agreements.

Leadership Principles and Tasks in Initiating and Convening

The principles of initiating and convening a collaborative effort include inspiring trust among stakeholders and encouraging their participation. The related leadership tasks implementing this principle include:

- Assessing the scope and timing of the collaboration;
- Identifying the interests and capable representatives based on that assessment, and gathering together the disparate voices needed to build consensus; and
- Building agreement among affected stakeholders to engage in good faith in the collaboration and consensus building.

Two challenges emerge: the tendency toward "flight to authority," and the pressures against an adequate assessment of feasibility and process design.

Conflict is a natural byproduct of adaptive work on water resource issues. Leaders often succumb to the temptation to seek a quick resolution without sufficient engagement with affected interests. In the *Ocklawaha River* restoration and de-authorization of the Cross-Florida Barge Canal, a premature legislative charge to implement a restoration plan led to impasse. Political leaders responded to what Heifetz calls the "flight to authority," pressing prematurely for legislative and technical fixes before scientists and stakeholders could research the problem and form a consensus (Heifetz and Laurie 1997).

Similarly, in the *Apalachicola* water case, the leadership from the Georgia, Alabama, Florida, and the U.S. Army Corps of Engineers failed to assemble a collaborative effort for the study initiated in 1991. In 1997, when President

Clinton signed the legislation creating the river basin compact, there had been little collaborative leadership during the previous six years. This wavering engagement and lack of continuity among policymakers from each state, and the inconsistent guidance for the technical staff, was exacerbated by the lack of engagement by stakeholders representing urban, agriculture, and rural interests, recreation, environment, shipping and navigation, and hydropower interests from upstream and downstream communities. This made the hard joint decisions difficult to conclude. As of mid-2005, the case remained at an impasse and was headed for the U.S. Supreme Court.

Heifetz argues that flight to authority inevitably fails to produce sustainable solutions because the situation is more than just a technical problem that leaders can solve, and because it keeps people from doing the hard work of learning together what adjustments and changes are needed. He advises leaders to put the work back on the stakeholders most directly affected—convene a balanced representation and help them come to a consensus (Heifetz and Laurie 1997). The *Everglades* case exemplifies this strategy. Governor Chiles had campaigned on finding a solution to the Everglades lawsuit. Initially he sought to exercise personal leadership to solve the problem. But he soon discovered the depth of frustration and complexity, and concluded that he needed to convene all those with a stake in Everglades restoration and give them an opportunity to become part of the solution.

Independent assessment of the process design and the likelihood of assembling key interests and producing an agreement is in the conflict resolution field considered a best practice (Dukes and Firehock 2001). The interconnections between ground and surface water systems prompted an innovative attempt to deal with the problems presented by nitrate pollution from agricultural operations along the *Suwannee* river basin. Facilitative leadership convened the Suwannee River Partnership and focused its efforts on the most problematic section of the river basin, the mid-Suwannee. However, the convening organizations, led by the executive director of the water management district, and the deputy commissioner of the state Department of Agriculture and Consumer Services, did not formally consider the range of interests in assessing the feasibility of building a consensus. While the executive director of the water management district was widely regarded as the "godfather" of the partnership, and successfully enlisted much of the agricultural community, the convening organizations were not able to make an offer that the environmental community couldn't refuse.

Instead, the environmentalists litigated against the rulemaking that made this partnership possible. This probably influenced EPA's later ruling, after much work among those in the partnership, that the partnership's proposed efforts in the mid-Suwannee did not provide reasonable assurances that the proposed control mechanisms would restore the mid-Suwannee watershed.

In *Everglades,* the process established a formal relationship between an intergovernmental Everglades task force representing federal, state, local, and tribal interests and a commission of stakeholders convened by Governor Chiles; this offered the best chance for consensus building and passage of the restoration plan (Jones 1997).

Leadership Tasks During the Collaboration

The leadership skills needed during a collaborative process are different than those needed to convene it. During collaboration, leaders must practice patience in the face of ambiguity and uncertainty, help to build trust among stakeholders in the legitimacy and efficiency of the process, and inspire the hard work of consensus building by offering meaning, vision, mission, and passion.

Leadership tasks that are especially important during collaboration process include:

- securing agreement among stakeholders on procedural guidelines, including a commitment to engage in good faith negotiations;
- participating as a peer problem solver with other participants, and sharing leadership functions as much as possible;
- building trust through interactions with stakeholders; and
- creating among stakeholders a sense of ownership of the problem and underscoring the importance of their work.

The tasks necessary to carry out these principles include organizing and sustaining a disciplined discussion, managing the rate of change and problem solving, orchestrating conflict, using adaptive management wisely, and safeguarding the integrity of the process.

Organizing and sustaining a disciplined discussion: This provides shared learning opportunities, including naming fears, managing expectations and collecting stories, definitions of issues and descriptions of problems (Heifetz 1997). Leadership often involves helping the stakeholders reframe separate narratives into questions, leading to the creation of shared narratives of the challenges faced. Stakeholders on the Governor's Commission for a Sustainable South Florida agreed, after 18 months of intense negotiations, that on its current course, the Everglades ecosystem was unsustainable. This helped reframe the story and the problem, and enabled the commission to develop consensus solutions.

Manage the rate of change and problem solving: Leaders must protect people by ensuring that changes do not come faster than buy in by those who must implement them. This is especially true in water resource conflicts where the collective histories have been long and difficult and the solutions will require prolonged commitment to adaptive management. Where leaders help to clarify shared visions and values, they help people adjust to new roles and responsibilities. Leaders must also assess what is happening from a wider perspective; what Heifetz calls "getting off the dance floor and onto the balcony." Collaborative leaders must have poise while exercising patience and persistence in maintaining tension without becoming overwhelmed: raising tough questions, fostering creative learning, and problem solving. This involves regulating disequilibrium, discomfort, impatience, and conflict—what Heifetz terms helping stakeholders "feel the pinch of reality without crumbling under the weight of the problem" (Heifetz and Laurie 1997).

Effectively managing conflict: People learn by engaging with a different view, not by staring into the mirror (Kanter 1999). Successful leaders manage the conflict between passionate and diverse points of view. Rather than shying away from conflict or suppressing it, they employ it as an engine of creativity and innovation (Heifetz 1994). Leaders must create an environment where conflict offers an opportunity for imagination, for learning about the risks, for consideration of multiple options, and for consensus building. The challenge is to develop structures and processes in which such conflict can be orchestrated productively. For example, once environmental interests found themselves at the table on Everglades restoration, they advanced the consensus-building effort, but at some cost to their relationships with environmental colleagues who believed that litigation might have served their interests better. Credible representatives of environmental interests in the Everglades participated directly in the consensus building on the conceptual plan and the USACE Restudy through the auspices of the Governor's Commission for a Sustainable South Florida. This transformed the earlier state and federal efforts into support of the more comprehensive consensus they had forged along with other stakeholder interests. They ensured that the environmental restoration goals of the Restudy received an appropriate level of attention (Jones 2002; Langton 1984)

Using adaptive management wisely: This includes testing options with "what ifs" in order to clarify assumptions, and gauging support for changes that may be needed as monitoring reveals new information. An early form of adaptive management in dispute resolution is the use of "prospective hindsight:" prompting stakeholders to develop criteria for judging a wise settlement and move substantive problem solving beyond the usual model of warring experts to a more cooperative approach. Susskind and Cruikshank (1987) suggest that when the parties focus on the issues and jointly utilize scientific and technical evidence regardless of who submitted it, a wiser outcome will result.

Safeguarding the integrity of the process: Leaders must work with their peers to develop the consensus-building format and guidelines, which includes charging representatives to explain to their constituents the process and the options under consideration. Leaders then need to protect the ongoing consensus-building and public-learning process from undue pressures from both above and below. They work to facilitate a credible, fair, and open process that carefully balances results, the process, and the relationships in preparing people for change. This means genuine participation in collective decisionmaking; this demonstrates respect, which in turn fosters trust. Leaders must ensure participation, build relationships, create value, and ensure accountability (Pettigrew and Fleischer 2002).

Leadership Tasks in Implementing a Consensus Outcome

> *It's the action, not the fruit of the action that's important. You have to do the right thing. It may not be in your power, may not be in your time, that there will be any fruit. But that doesn't mean you stop doing the right thing. You may never know the results which come from your action. But if you do nothing there will be no result.*
> Mahatma Ghandi

Before the Governor's Commission for a Sustainable South Florida issued its first consensus report in 1995, chairman Richard Pettigrew read into the record and included in the letter to the governor this quote from Gandhi. He believed that it best expressed his own view of the quality of leadership needed for actions whose fruits would not be known immediately but would benefit future generations of South Floridians.

In implementing a consensus outcome, perhaps the greatest challenge in water resource conflicts is to hold accountable those implementing the results, when this may take years, even decades, to accomplish. This is especially true where adaptive management is an integral part of the plan.

Building public and bipartisan support for implementation by established authorities, such as legislative bodies, is a leadership task of those bringing forth the solution. Sometimes, as was the case in the Everglades restoration case, a new leader who was not part of the consensus building must implement the solution. While Governor Chiles had supported the consensus-building efforts of the Commission over five years, and sought settlement of the Everglades dispute since the beginning of his term in 1990, it was his successor Governor Bush who in 2000 elicited unanimous support in the state legislature for the consensus restoration plan, and helped to secure congressional support for the 50–50 partnership. During most of the 1990s Democratic Senator Bob Graham and Republican Senator Connie Mack, along with the Florida congressional delegation, were able to work as a team in support of Everglades restoration plans and funding.

Leaders also need to acknowledge that forecasts often fall short (Kanter 1999). Plans are based on experience and assumptions. It is difficult to predict how long an innovative process will take, what will be learned in the course of implementation, or how much it will cost. Leaders must be prepared for serious deviations from plans, and discuss how they will be dealt with. Dealing with the results and the potential new conflicts that adaptive management entails has received less attention than it deserves (Lee 1993).

Building an Institutional Commitment to Collaborative Leadership

How can we do a better job of developing collaborative leaders? Continued reliance on an ad hoc call to leadership by unprepared individuals leaves too much to chance. Statewide leadership programs, such as Leadership Florida (2004) and local leadership programs have begun to concentrate on collaboration skills and engagement with others in policy development process, and less on traditional policy issue briefing. Reliance on ad hoc leadership mechanisms should be supplemented by a greater institutional commitment. These public, private, and nonprofit organizations should press for a collaborative leadership curriculum that can teach these skills.

Two national organizations devoted to integrated governance at the state and local levels through collaboration among citizen, public, private and nonprofit, and philanthropic sectors, have developed a new leadership initiative. The Policy Consensus Initiative and the National Policy Consensus Center, both based at

Portland State University, have launched the Collaborative Governance Network of leaders with convening and consensus-building skills. The centers have designed an initiative drawing on Dee Hock's work on chaotic, complex adaptive systems and organizations (Hock 1999). Complex systems for public policy and problem solving are constantly evolving, and the patterns that govern these systems are emergent. To succeed in this environment, Hock has suggested that institutions need to identify and exploit emerging patterns, support careful reflection and exchange of information on what is being learned, and sponsor experimentation based on this public learning.

The leadership development program is bringing together state leaders—governors, legislatures, and their staffs—as well as universities and other funders to develop and use new collaborative governance strategies and principles, and to evaluate and disseminate the lessons learned. A key strategy is enlisting political leaders to lead the collaborative processes and spread the word to other leaders regarding successful innovations, principles, and models for governance. In convening a collaborative process, these organizations suggest that leaders:

- provide a neutral, transparent setting;
- include all interests willing to contribute; and
- ensure that government and others implement the solutions together (Policy Consensus Initiative 2004).

Some states have created new leadership training programs called Natural Resources Leadership Institutes. The first such program, supported by the Kellogg Foundation, was developed in North Carolina in the mid-1990s. In Florida, the University of Florida's Cooperative Extension Service, as part of the Institute of Food and Agricultural Sciences and the Florida Conflict Resolution Consortium housed at Florida State University, joined to create the Florida Natural Resources Leadership Institute (FNRLI) in 1998. They have worked together for the past seven years to help rising leaders in industry, government, and the environmental community enhance their skills for consensus building and managing conflict over natural resource issues.

Each FNRLI class is composed of approximately 25–30 "fellows" representing these three sectors. They participate in a year-long program that meets around the state to engage environmental challenges firsthand. They study personal and group leadership skills, communication skills, dispute resolution and consensus-building techniques, and environmental law and policy. Fellows often enter the class expecting to find adversaries holding vastly different values and goals, but find through the leadership program that there is substantial agreement about the underlying values and importance of natural resources in Florida as well as the role that collaborative problem solving can play.

These programs are embryonic; much more could be done to refine their curricula and connect them with those developing new adaptive collaborative governance concepts and approaches.

The Role of Collaborative Leadership in Public Learning

This paper explored the need for leadership strategies and skills to make the most of public learning and bring forth solutions on water conflicts. Collaborative leadership will be needed to apply adaptive governance to building consensus on water resource solutions, and as a strategy to implement that solution. The growing leadership literature on the adaptive challenges of troublesome public issues offers some helpful framing concepts. Applying adaptive management to the institutional and political structures that govern water resources will require a creative rethinking of governance structures, leadership strategies, and institutions that can train new leaders.

The *Everglades* case suggested that the quality of leadership exhibited by many who have been involved in ecosystem restoration in South Florida may be the most important factor. "A profile of these leaders suggests similarities in temperament, values and skills, and above all, a strong orientation toward collaboration and consensus" (Langton and Salt 2003, *249*). The very nature of a large, complex ecosystem restoration effort demands these qualities of leadership. Further, it suggests a reinforcing effect among leaders of equal status and the successful leadership of their predecessors.

Collaboration on water resource projects and permits will be a continuing challenge for leaders and stakeholders. There is a powerful and lasting legacy of leadership that was based on the command-and-control leadership model. Learning a new leadership language that better fits the complex and adaptive world of water resources issues and the role of public learning in these initiatives must be part of the adaptive governance model. This leadership challenge and way of working is the responsibility of not only those in positions of power, but also representatives of the disparate interests, as well as scientists and managers who are brought together for consensus building.

CHAPTER 17

Public Learning and Grassroots Cooperation

Mark Lubell

PUBLIC LEARNING IS an important basis for adaptive governance. Public learning includes both increasing knowledge about the possible outcomes of different policy and behavioral choices, and also changing views about the legitimacy of decision processes and behavioral restrictions. This chapter focuses on two aspects of public learning: the factors that determine stakeholders' views on the effectiveness of water management policies, and their participation in water management activities. These attitudes and behaviors are both necessary to lay the groundwork for broader cooperation and policy implementation.

"Grassroots stakeholders" are the people who actually consume natural resources—the fishers, the farmers, the water diverters, the loggers, and other species of what Ostrom (1990) calls "appropriators." The discussion of public learning in the Introduction correctly emphasizes the role of grassroots stakeholders. Their resource decisions are the immediate cause of most environmental problems—how much they take from the environment, using what technologies, and the types and nature of substances they put back into the environment. Hence, the success of adaptive governance depends on public learning that ultimately leads to environmentally sustainable changes in the behavior of grassroots stakeholders.

However, most of the research on adaptive governance has centered on new decision processes, institutional rules, and interactions among policy elites from the private and public sectors. This bias extends to the case studies presented in this book. For example, the discussion of the *Apalachicola* tri-state water conflict mainly examines negotiations among three states, interest groups, and relevant federal authorities. The *Everglades* case study emphasizes the variety of intergovernmental planning forums that eventually produced the Comprehensive Everglades Restoration Plan. If these policy elites can agree on a set of rules for governing water resources, and can change those rules in response to public learning, then adaptive governance may succeed.

Yet grassroots stakeholders are critical factors in both conflicts. In the Everglades Agricultural Area, individual farmers are required by state law to obtain a water discharge permit from the South Florida Water Management District. The permits contain a best management practices (BMP) implementation plan and on-farm water quality monitoring plan. The regulatory nature of the Everglades BMP program makes public learning a matter of learning what is required to comply with the law, and the consequences of noncompliance. The water supply issues in the Apalachicola conflict may eventually involve recommendations for limiting urban water use in the Atlanta metropolitan area and improving the efficiency of downstream farm irrigation systems. Increasing the efficiency of urban and agricultural water use entails behavioral change and public support for the legitimacy of the policies. These issues are ubiquitous in the American West and will become more common in the East as water demand increases.

To remedy the lack of academic attention to the behavior of grassroots stakeholders, this chapter places them directly under the social science microscope. In particular, I examine several theoretical frameworks that claim to explain the behavioral underpinnings of cooperation. To test their validity, I have surveyed farmers in the Suwannee River Partnership. To reiterate, the goal of the Suwannee Partnership is the collaborative implementation of water management plans and best management practices by Suwannee farmers. Hence, farmers are the most important grassroots stakeholders in the Suwannee basin.

Why should we believe an analysis of Suwannee Partnership farmers can produce valid conclusions for other water governance problems in Florida and elsewhere? First, the Partnership is an example of collaborative policy, which is the distinctive method of decisionmaking in many of the other case studies, and also one of the main alternatives to coercive regulations. While adaptive governance does not necessarily imply collaboration—there are certainly circumstances where regulation does work—the complexity of watershed problems provides a niche for the evolution of collaborative institutions. Second, agriculture is involved with nearly every collaborative partnership focused on watershed management. Third, while Suwannee farmers are not necessarily representative of other Florida or United States agricultural communities, there are enough similarities across agricultural communities (e.g., distrust of regulatory agencies) to suggest that the conclusions reached here will apply elsewhere. Fourth, to the extent other water conflicts create similar problems of cooperation and public learning for other types of grassroots stakeholders (e.g., residential water uses in Atlanta), the same basic theoretical frameworks should be applicable. At the very least, researchers and practitioners can start asking the right questions about the grassroots view of adaptive governance, and begin to think about theoretical explanations for any differences between the conclusions of this analysis and other research projects.

Explaining Farmer Cooperation

Cooperation ultimately refers to farmers agreeing to implement best management practices on their farms. I divide cooperation into two essential elements:

perceptions of the effectiveness of practices recommended by the partnership, and farmer participation in partnership activities. If farmers do not think partnership policies are effective, they are much less likely to cooperate, and more likely to engage in political strategies designed to weaken those policies. Participation is the behavioral manifestation of cooperation, and allows farmers to learn about sustainable practices and apply them according to water management plans.

I examine three main theoretical perspectives to explain farmer cooperation: rational choice, social capital, and social values. In my interviews with Suwannee farmers and policy elites, they mentioned elements from all three of these perspectives as explanations for farmer cooperation. Even within the agricultural community, there is a diversity of opinion. Each perspective reflects well developed social science theories, and the empirical analysis suggests that multiple factors influence cooperation.

Rational Choice

The rational choice perspective on farmer cooperation is largely built on economic models that posit humans always choose behaviors perceived to have the highest benefit–cost ratio. Especially when farmers do not agree that a water quality problem exists, they tend to resist any type of pollution control policy that they believe will increase their production costs. They are more likely to accept government policies that provide financial incentives, which is why most BMP programs feature cost-sharing arrangements. For example, the Suwannee Partnership coordinates cost-share programs from United States Department of Agriculture's (USDA) Environmental Quality Incentive Program (EQIP) and Public Law 566.

BMP programs also emphasize potential increases in production efficiency that come from BMP implementation. BMP often reflects a philosophy of reducing resource inputs: water used per acre, pounds of fertilizer, or amount of pesticides. In many cases, farmers report being surprised at how much money they save once they start using input-reducing BMPs.

Another important economic consideration is the threat of future regulations, and the possibility that voluntary conservation could provide regulatory relief. All over the country, the hammer of future regulation is an important motivation for current collaboration. In the Suwannee River area, the regulatory hammer is the required development of a Total Maximum Daily Load (TMDL) for nitrates and a plan to meet the TMDL requirements if the river continues to violate water quality standards. If the Partnership can demonstrate with "reasonable assurance" to the U.S. Environmental Protection Agency (EPA) and the Florida Department of Environmental Protection (FDEP) that Partnership activities will eventually improve water quality, then the TMDL process can be deferred.

The rational choice approach is closely related to the diffusion-innovation model developed by Ted Napier and his colleagues (Napier and Camboni 1988; Napier et al. 1988a, 1988b; Napier and Tucker 2001). The diffusion-innovation model focuses on farmers' evaluations of the benefits and costs of adopting conservation or general agricultural practices used by other farmers. It also looks at the economic constraints imposed by farm size and economic health, and generally argues larger and more economically viable farms can better assume the risks of new behaviors. Interestingly, despite a series of sound empirical studies, Napier and his colleagues conclude the diffusion-innovation model is not a sufficient explanation of farmer behavior, and call for alternative theories.

Social Capital

The social capital perspective views voluntary partnerships as a collective action problem. Farmers have a collective interest in reducing water pollution, either because they value clean water resources, or because clean water will prevent further regulatory action. However, the defining characteristic of nonpoint-source pollution from agricultural runoff is the difficulty of identifying which farms are contributing contaminants to the river. Hence, one farmer's conservation practices will only have a small influence on the overall water quality in the watershed. A strategic farmer prefers to free-ride on the conservation practices of others, avoiding the individual cost of BMP implementation and enjoying the benefits provided by his neighbors (Hardin 1982; Olson 1965; Ostrom 1990, 1998). The free-riding strategy is often coupled with a denial that runoff from the farmer's land is contributing to the water quality problems. If all farmers follow this strategy, then the rate of BMP implementation is much lower than required for economic efficiency or environmental sustainability.

Solutions to this problem focus on the repeated interactions among farmers in multiple social arenas, and also the repeated interactions between farmers and the government officials who are involved with BMP programs. Repeated interactions allow for the development of "social capital," in particular norms of reciprocity, trust, and networks of civic engagement (Coleman 1990; Putnam 1993; Schneider et al. 2003; Ostrom 1998). Norms of reciprocity will create the perception that the implementation of BMP by one farmer will increase the probability that other farmers will reciprocate and implement BMP on their farms. Expectations of reciprocity are a key to successful collective action (Finkel and Muller 1998; Finkel et al. 1989); people are more likely cooperate when they believe that others will reciprocate. Hence, farmers who believe that implementing BMPs on their farms will lead other farmers to do the same are more likely to participate and view BMP polices as effective.

Trust also reflects beliefs that other parties in a social exchange relationship will fulfill their commitments. Trust between government officials and farmers is particularly important, because government officials are asking farmers to implement BMPs in exchange for promises of cost-share money, accurate information, and perhaps regulatory relief. If farmers do not trust government officials to live up to these commitments, they are less likely to cooperate. Overall, the development of social capital helps facilitate long-term cooperation in situations where short-term cooperation is predicted to fail.

It should be noted that the social capital perspective is not at all contradictory to the economic perspective, but it does emphasize different variables. The social capital perspective assumes that cooperation has long-term economic benefits either from improving water quality or avoiding regulatory intervention. In other words, cooperation is in the long-term economic self-interest of farmers.

These benefits can only be achieved if enough social capital develops to support cooperation over time. The development of social capital may outweigh the more short-term economic benefits of cost-share provisions or increases in farm efficiency from BMP implementation.

Social Values

Like all people, farmers have social values about the environment, government, and society. According to Sabatier and Jenkin-Smith's (1993) Advocacy Coalition Framework (AC), these social values are integrated into fairly cohesive belief systems, where more fundamental "policy-core beliefs" constrain the formation of more immediate "secondary beliefs" about attitude objects in a policy subsystem (Hurwitz and Peffley 1987). The ACF argues that people tend to discount information about an attitude object that is inconsistent with their core beliefs, and accept consistent information. To the extent the collaborative spirit of the Suwannee Partnership and the concept of best management practices are consistent with farmers core beliefs, they should view the BMP program as effective (Lubell 2003). While the AC does not explicitly consider the effects of core beliefs on behavior, one can extend their logic and hypothesize that core beliefs influence the behaviors associated with a particular attitude object.

Perhaps the two most studied core beliefs in the AC literature are views on the proper role of government in regulation of private rights, and environmental ideology. The former, which I will label *economic conservatism*, ranges from people who think the government should heavily regulate economic activity to prevent undesirable externalities, to people who believe government should not interfere in economic activities. *Environmentalism* measures the extent to which people value environmental amenities, or feel that human activities threaten environmental values. In the context of collaborative partnerships and voluntary BMP programs, which often try to resolve conflict between economic and environmental interests, there should be a positive relationship between both economic conservatism and environmentalism, and perceptions of BMP effectiveness and participation.

I also examine the core beliefs of *land stewardship* and *policy inclusiveness*. The agricultural community hangs many policy positions on the hook of land stewardship, arguing that farmers naturally care about their land because their land is the basis for their livelihood. Hence, government and environmentalists should trust the agricultural community to voluntarily protect environmental values. The concept of land stewardship is not just a figment of agricultural lobbyists' imagination. A comment by one survey respondent is representative: "We as land owners are morally bound to utilize BMP's and practice stewardship of our land. Those of us who were privileged to grow up on a farm have a greater appreciation of the land and its importance to mankind than those who buy land to escape the urban areas." Land stewardship has a long history in the ideology of the agricultural community, with roots going back to Jeffersonian values of small-scale agriculture. Hence, farmers who subscribe to the ideology of land stewardship are more likely to view BMPs as effective and participate in the program.

Policy inclusiveness is the belief that public participation in policy decisions should be maximized. Inclusiveness is one of the basic tenets of collaborative policy, which generally seeks to forge consensus among all stakeholders. Previous research on collaborative policy found that policy elites committed to the concept of inclusiveness are more likely to believe collaborative partnerships are effective (Lubell 2003). Expanding this hypothesis to the grassroots, I expect farmers committed to inclusiveness are more likely to view BMPs as effective and participate in the Partnership.

The Suwannee River Partnership Farmer Survey

In this section, I briefly summarize the results of a mail survey sent in Winter 2002–2003 to all farmers in the Suwannee Basin who were potential targets for the Suwannee River Partnership. The survey measured their participation in the Partnership, their views about BMP effectiveness, and a variety of independent variables associated with the three perspectives on farmer cooperation discussed above. From an effective sample size of 299 farmers, 83 completed the survey, for a total response rate of 28 percent. Comparison of respondent characteristics to the Census of Agriculture suggests that the sample is biased towards larger, richer, and more politically involved farmers. Twenty-five percent of the respondents are beef cattle or dairy producers, 47 percent are poultry producers, and 27 percent are some type of row crop producer. The row crop producers are probably under-represented, because the Suwannee Partnership had an early focus on the better-organized dairy and poultry producers.

Analysis Results: Bivariate Correlations

Table 17-1 presents bivariate correlations between two indicators of cooperation—*perceived BMP effectiveness* and *participation*—and the key concepts identified by each of the theoretical perspectives discussed above. Perceived BMP effectiveness is a seven-point Likert scale that measures whether the farmer believes BMP will improve the environmental health of the Suwannee River, including farm-level outcomes. Participation is a count of the number of partnership activities in which the farmer has taken part. The survey offers eight possible activities, ranging from meeting with Partnership representatives to development of an on-farm water management plan. Attending meetings and talking with Partnership representatives are the most frequent activities, and attending BMP training sessions the least. A positive and significant correlation (Pearson's $r = .30, p < .05$) between participation and perceived effectiveness reinforces their meaning as two aspects of farmer cooperation.

Survey questions also measured each of the key independent variables indicated in the table (Lubell 2004). In another analysis (Lubell 2004), I test the robustness of these correlations using multivariate regression. With only a couple of exceptions (farm economic status and economic conservatism become insignificant in the participation regression model; environmentalism becomes insignificant in both regression models) the significant correlations remain signif-

Table 17-1. *Factors Affecting Perceived Effectiveness and Participation*

	Perceived BMP effectiveness	Partnership participation
Economic Perspective		
Farming Efficiency	.44*	.06
Cost-Share Programs	.35*	.11
Regulatory Relief	.16	.18
Farm Size	-.04	.25*
Farm Economic Status	.15	-.21*
Social Capital Perspective		
Expected Reciprocity	.44*	.32*
Local Government Trust	-.02	.24*
Regulatory Agency Trust	-.11	.07
Farm Industry Trust	-.07	.08
Social Value Perspective		
Economic Conservatism	-.01	-.27*
Environmentalism	.23*	.21*
Stewardship	.25*	.08
Inclusiveness	.32*	.01

Notes: Entries in cells are Pearson's correlation coefficients.
*Reject null hypothesis of correlation coefficient = 0; $p < .10$.

icant in the regression analyses, and those that become weaker in the regression analysis would probably remain valid if the sample were larger.

Two economic variables, farming efficiency and cost-share programs, are positively correlated with perceived effectiveness. Farmers clearly prefer management practices that do not increase their production costs. Self-assessed farm size (higher values mean respondent views their farm as larger than average) is positively correlated and farm economic status (higher values mean the respondent thinks they are much worse off than a year ago) is negatively correlated with participation; the Partnership appears to be most attractive to larger and richer farms. Surprisingly, the expectation that participation in the partnership will reduce the threat of future regulation does not affect perceived effectiveness or cooperation.

The social capital variable of expected reciprocity is positively correlated with, and is the largest coefficient for, both BMP effectiveness and participation. Expected reciprocity explains 19 percent ($r^2 = .19$) of the variation in BMP effectiveness, and 10 percent ($r^2 = .10$) of the variation in participation. The only other significant social capital variable is local government trust, which is positively correlated with participation. Social capital, especially expected reciprocity, plays a critical role for both perceived effectiveness and participation. But the added importance of local government trust suggests social capital is more important for participation.

The social value variables have a variety of interesting relationships. As predicted, environmentalism is positively correlated with both BMP effectiveness and participation. Farmers that express a commitment to environmental values are more likely to cooperate. Both stewardship and inclusiveness are positively correlated with perceptions of effectiveness, but not with participation. Counter to the hypothesis that conservatives will positively view collaborative policies, economic conservatism is negatively correlated with participation. This suggests that collaborative policy still reflects some of the traditional conflicts between economic and environmental values.

Three generalizations stand out from this analysis. First, and most important, is that *expectations of reciprocity are a driving force in farmer cooperation*. Given the collective action problem associated with improving water quality through BMP implementation, there are no benefits from cooperation by a single farmer unless that farmer believes his behavior will affect others. Cascades of reciprocal behavior will increase the probability of providing the public good of improved water quality, making the benefits of cooperation appear to outweigh the costs. These results are entirely consistent with the collective interest model of cooperation proposed by Finkel et al. (1989).

Second, *collaborative partnerships appear to be a two-part process*. One part consists of getting people involved in partnership activities that revolve around interactions with government officials and other policy elites. Policy practitioners often talk about this as "buying in" to a process or "getting the stakeholders at the table." Social capital variables are very important for this stage, including trust in the local government officials. But once farmers buy into the collaborative process, which ostensibly means deciding that participation of some type is a good thing, they become more concerned with the economics of BMP implementation. In other words, the decision to cooperate in some fashion is made on the basis of social capital, but once that decision is made, farmers pay close attention to selecting the least costly form of cooperation available.

Lastly, *there is a disjunction between the expression of attitudes like perceived BMP effectiveness and cooperative behavior*. Policy-oriented beliefs, in particular social values, provide more explanatory power for BMP effectiveness than for participation. Conversely, structural characteristics like farm size and economic viability, and the quality of the social relationships among all stakeholders, have a greater influence on participation. The correlations in Table 17-1 show this disjuncture, and the weakness of environmental values in the regression models for participation reinforce the point. Interestingly, these observations resemble the disjuncture between attitudes and behavior seen in more general analyses of environmental activism (Lubell 2002b). While the average citizen's expressed support for environmental policy is heavily influenced by social values, their actual activism behaviors are more influenced by demographic characteristics like education, age, race, and environmental knowledge, which affect the costs of environmental activism.

Implications for Adaptive Governance

The results of the analysis suggest that the view from the grassroots has important implications for several aspects of adaptive governance. Most importantly, my central argument is that one important element of adaptive governance is public learning and cooperation from grassroots stakeholders. Cooperation entails a behavioral component of participation in a collaborative partnership or other watershed management activity, and attitudinal support for particular sets

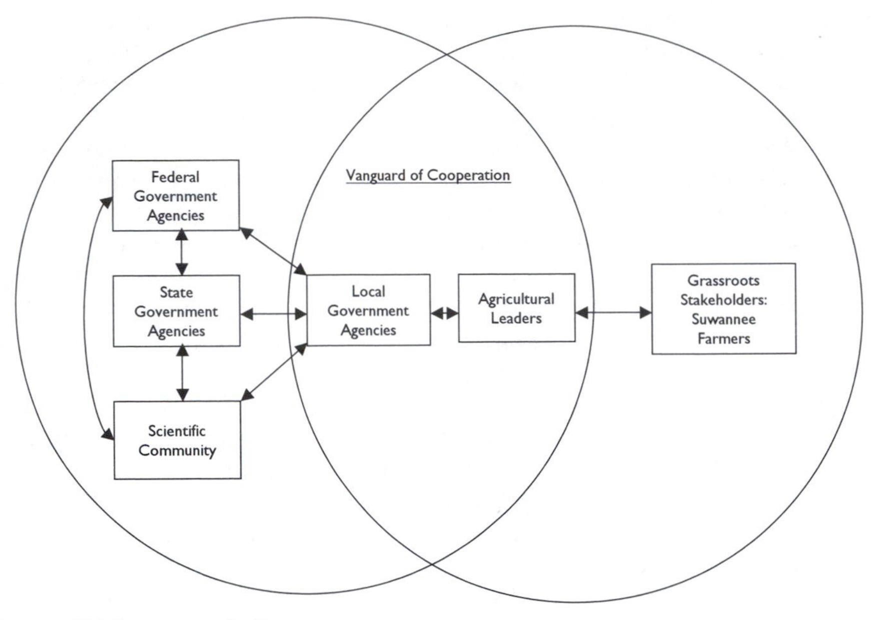

Figure 17-1. *Network Structure*

of policies. Positive attitudes and behaviors are both necessary, but not sufficient, to produce cooperation in the context of a collaborative partnership. For farmers, participation is driven largely by expectations of reciprocity from other farmers and trust in local government agencies. Once participation is secured, cost considerations and social values become more important in how they view the effectiveness of various policy options.

What do these results say about the challenge of process design? At least in the early stages, adaptive governance processes should strive to create a *vanguard of cooperation* consisting of a core policy network of trusted government officials and leaders representing relevant groups of grassroots stakeholders. The resulting network structure is sketched as a Venn diagram in Figure 17-1, where double-headed arrows represent reciprocal, trust-based relationships between categories of policy stakeholders. In the Suwannee, the vanguard consists of local government agencies and agricultural leaders, including organized agricultural producer groups and cooperatives. The vanguard becomes a nexus for public learning by spanning two important peripheral networks. First, the local government agencies tap into the wider intergovernmental network that may include scientists or less trusted regulatory agencies like EPA and FDEP. The resulting intergovernmental coordination would integrate existing legislative and judicial rules, and also acquire essential information. Second, the grassroots leaders tap into the network that they represent; they can deliver information, use persuasion to change preferences, and overall encourage cooperation with adaptive governance rules. The interactions in the vanguard of cooperation essentially mediate between two sets of actors that are often in conflict. In the language of policy network theory, the vanguard spans a structural hole and enhances social capital (Burt 1992). In the language of evolutionary theories of cooperation, the vanguard serves as a cluster of actors playing reciprocal strategies that can help cooperation become more widespread in a population (Axelrod 1984). Hence, over time, a successful adaptive governance network would develop cooperative relationships between actors who are often in conflict, such as farmers, environmentalists, and regulatory agencies.

These results also pertain to the challenge of problem responsiveness, in particular equity, efficiency, and dynamic adaptability. With regards to equity, partnership (and survey) participation appears skewed towards larger and more economically viable farms. Hence, many smaller farms may not contribute an equal amount of effort to partnership activities. If all farms contribute a fairly equal amount of nitrates to the Suwannee basin, then the larger farms might complain about an unfair distribution of costs. However, they may also be willing to tolerate free-riding from smaller farms in order to secure the benefits of cooperation—Olson (1965) might call the larger farms a privileged group. But the distribution of effort might seem more equitable if larger farms also contribute a disproportionate amount of nitrates to the river. Evaluating the equity of equal versus proportionate levels of cooperation is a venerable dilemma in distributive justice (Hardin 1982, *90–100*), and may even be impossible to solve even with perfect information about the nitrate runoff from every farm.

Efficiency evaluations must take into account the appropriate balance between private conservation expenditures and public cost-share money. Clearly, cost-share money is an important motivation when farmers choose among alternative BMP strategies. Some environmentalists would argue that farmers should pay for all of the conservation measures, while some conservatives would argue the public should pay for everything. Striking a balance between these two extremes must include an evaluation of how much conservation is achieved for each dollar of cost-share money. That evaluation should include not only a technical assessment of BMP effectiveness, but also the increased motivation for BMP adoption that cost-share money provides.

Lastly, adaptive governance requires monitoring the effectiveness of policy implementation for solving water problems. Adaptive governance should apply the environmental monitoring protocols of adaptive *management* to human behaviors and attitudes, tracking their change over time and adjusting policies in light of results. In the context of human–environment interactions, this monitoring must take into account the uncertain causal linkages between attitudes, behaviors, and environmental outcomes. If one observes positive attitudes towards BMP, there is no guarantee that cooperative behaviors will ensue. Similarly, if one observes cooperative behavior, there is no guarantee that environmental outcomes will improve. The cooperative behaviors, for example the types of BMP implemented on the farms, may not have a large effect on environmental outcomes.

Interestingly, the "reasonable assurance" concept embedded in the Florida water code reflects this uncertainty as applied to the Suwannee. Currently, the reasonable assurance documentation is based on estimates of farmer participation in the partnership, supplemented with scientific studies that measure the pollution reduction effectiveness of specific BMPs. Monitoring farmer participation and adoption of recommended practices is a sufficient short-term strategy, especially in the case of a groundwater problem like the Suwannee where the effects of behavioral changes may not be seen for decades. However, for the Suwannee Partnership to become a sustainable experiment in adaptive governance, reasonable assurance must be directed to watershed outcomes such as monitoring ambient surface water and groundwater quality.

CHAPTER 18

Putting Science in Its Place

Connie P. Ozawa

UNTIL RECENTLY, in many urban areas of the eastern United States water was only rarely disputed publicly. Decisions were left to water service providers and engineers, with the blessing of elected officials when necessary. Urban residents hardly noticed the activities of their service providers, except when a new, expensive water or sewage project showed up as a rate increase on their monthly bills. These eight cases of water conflict in Florida foreshadow change in an arena of public decisionmaking long taken for granted. What can we learn from them that will help us to address future challenges?

Water conflicts invite a simplistic urge to use the "best" science to make decisions and settle disputes. However, rising demands by an increasingly diverse group of users—not only residential consumers, industry, and agriculture, but also recreationists and advocates of ecosystem protection and restoration—have intensified competition and complexity, and added new parties to the fray. What qualifies as the best science is now often contested, and must be accepted by a wider array of parties than ever: not only resource managers at local, state, and federal agencies, but also elected officials, professional representatives from user groups, and ordinary citizens.

The Florida water cases illustrate the failures and successes of integrating scientific and technical information into policymaking. They prove that the way in which science is introduced into decisionmaking does matter. They suggest that the use of scientific and technical analysis can aggravate conflict or narrow differences. They also suggest that water conflicts do not always revolve around disputed science. How science and other types of information such as technological feasibility studies, economic cost estimates, or "local knowledge" are integrated into decisionmaking cannot be analyzed in isolation from other facets of the process. It is neither what information is accessed nor how information is handled in the process that makes or breaks a decision. Parties agree when their political interests are sufficiently satisfied and their faith in the legitimacy of the process is intact.

This chapter considers challenges of addressing science in public decisionmaking. I suggest that despite the widespread recognition of the political content of scientific and technical information and analyses, we only partially understand the changes in behaviors and institutions necessary to create a system of collective decisionmaking to support adaptive governance. The Florida water conflicts exhibit a variety of uses of science. Some illustrate the persistence of missed opportunities for a more productive employment of information. Others demonstrate constructive techniques. This discussion addresses political representation, scientific learning, public learning, problem responsiveness, and decision process design. I first review the nature of scientific work and existing barriers to more effective integration. Next I discuss ways to incorporate science that appear to hold some promise for sustainable decisionmaking. Techniques and procedures alone, however, are not sufficient. Returning to the proposition that even the best science will not end disputes, I close with four themes central to creating a system of adaptive governance.

Challenges of Science in Public Decisionmaking

In water conflicts, as in other science-intensive situations, we know that our knowledge is incomplete. We know this uncertainty will persist. Nonetheless, we continue to create expectations that a best science can be objectively identified and, in many cases, we simply choose to ignore the meaning of "the social construction of knowledge." Not surprisingly, our efforts to reach decisions often are derailed. Adler et al. (2001) have identified 23 obstacles to science-intensive negotiations. They call these "rockslides in the road" toward agreement. The 23 "rocks" can be clustered into three sets of challenges, according to the ease of their solution, and these case studies provide useful examples.

The first set of rocks obstructs access to information, expertise, and the quantity and quality of existing of data. Parties less comfortable with scientific and technical issues may feel that other parties are manipulating technical information and analyses to justify their own policies. In defense, they may try to deflect discussions away from technical issues. Instead, they may find themselves marginalized in discussions and their political demands muted and unattended, only to resurface further on in the process. Reservoir supporters in *Ocklawaha*, for example, remained uninvolved in the agency-directed technical studies dominated by pro-restoration scientists only to emerge later in the legislature with enough clout to prevent funding; their success was unrelated to the science involved. Data gaps or other known deficiencies further impede discussions, inducing some parties to advocate deferring a decision, although delay itself can benefit certain groups at the expense of others, and can therefore be interpreted as a contentious act.

A second set of rocks emerges when some parties will not accept scientific and technical information. Students of information and decisionmaking have discussed for several decades now the extent to which science is socially constructed (Andrews 2002; Fischer 2000; Kuhn 1962; Latour 1979; Ozawa 1991; Schraeder-Frechette 1993). We acknowledge the reliance of formal analysis on untested assumptions in order to speculate about future events and large ecosystems, necessary simplifications made in the course of creating models and identifying variables for analysis, limitations in the spatial and temporal expanse of data, and other shortcomings of reductionist approaches. Moreover, as Bisbal has pointed out, information and analysis important to environmental management include a range of scientific and quasi-scientific methodologies that vary markedly in their similarity to formal science (Bisbal 2002). Some forms of "scientific information" are little more than expert judgment, valuable in their own right, but not replicable to the same degree as laboratory experiments, for example. Despite this variability, we do not insist on revealing discretionary components of knowledge in our public policy debates. We continue to design our decisionmaking processes in ways that ignore embedded subjective judgments that can have important political implications.

As a result, parties quite sensibly view studies put forth by policy advocates and purchased information skeptically, and are wary of others' strategic use of information. In *Aquifer Storage*, for example, environmentalists resisted the experts' projections about the fate of water injected underground. In other cases, the scientists' labor may be directed at questions not of primary concern to the decisionmakers. *Apalachicola* provides an excellent example: decisionmakers from Georgia continued to ignore the pleas by downstream Florida biologists concerned more about habitat impact than water flow. While scientists often labor meticulously over their studies in order to arrive at their findings, the presentation of a "black box" may be sufficient reason for parties to discount their work. Although outright rejection of their work without an opportunity to explore its relevance may frustrate the experts, indifference to their findings may be a logical response to analysis performed without consultation. Challenges to acceptability are more complicated to address than lack of access to data or expertise, and they inevitably arise.

A third set of rocks springs from the incompatibility of the scientific enterprise in its purest form with the practical demands of public decisionmaking. This set includes accommodating uncertainty and instances of what Alvin Weinberg identified as "trans-scientific" questions, a class of questions important for policy, but impossible for science to answer (Weinberg 1972). Predicting underground water flows as in *Aquifer Storage*, water quantity and quality needs of ecological systems during periods of climatic stress as in *Apalachicola*, and the unending quest to reveal the inner workings of the *Everglades* are all examples of trans-scientific questions. Difficulties in decisionmaking surface because science as a field has high tolerance for ambiguity—"Competing versions of scientifically derived 'truth' can, and often do, coexist" (Ozawa 1991)—whereas policymakers and the electorate demand more clarity and conclusiveness.

A concrete example of the dissonance between science and public decisionmaking is reaction to surprise. Scientists are not put off balance by the unexpected. Instead, they welcome it; new information may hold the key to unlocking more pieces of the scientific puzzle. "In terms of experiments...surprising results are legitimate, rather than signs of failure" (Lee 1993). In contrast, surprise in the public arena arouses suspicion or contempt. Decisionmakers dread having to admit that they had based their policy positions on analyses later found to be inaccurate. The unexpected is regarded as a threat to existing negotiated agreements. Individuals and organizations are viewed suspiciously if they modify their understanding of technical dimensions. As a result, although science is most appropriately conceived as a process in which surprise is acceptable, existing policymaking dynamics create expectations of predictability that are both unrealistic and antithetical to the nature of discovery itself.

The Florida cases show that despite our enlightened awareness about the nature of knowledge acquisition, public conflicts are littered with many of these "rocks," and our decisionmaking processes do not deal with them adequately.

Guidelines for Decision Process Design

What would a system of decisionmaking more congruent with our understanding of the social nature of knowledge require? The Florida cases suggest some changes. Any attempt to modify the dynamics of relationships and interactions among competitors for scarce resources must take seriously existing legal rights and protections and must be undertaken cautiously. Nonetheless, the list in Table 18-1 is offered as a start toward transforming our thinking and behavior in the use of science. It gives examples of practices that have been or might be implemented in these Florida water conflicts and elsewhere.

Disseminate information regularly. The importance of disseminating information is popularly recognized, and mechanisms for doing so are well developed. Public outreach and education is viewed as a critical element of efforts to nurture a shared understanding of public issues and garner public support. Disclosing the basis for decisions is integral to the accountability of elected officials and admin-

Table 18-1. *Devices for Improving the Use of Science*

- Regular dissemination of information in highly accessible formats
- Workshops, panels, and other opportunities to disclose and explain discretionary elements of research and analysis
- Sharing technical expertise
- Joint fact-finding
- Public statements by resource managers and elected officials acknowledging the incomplete state of knowledge and the role of surprise
- Public explanations of discrepancies between or among different expert advice by resource managers
- Short-term decisions with provisions for review
- Ongoing monitoring and data collection
- Risk-sharing (benefits, costs)

istrative agencies. Not only must information be available to all, but a minimal level of competency and comprehension must be assured if democratic ideals of representation and participation are to be achieved (Marshall and Ozawa 2004; Ozawa 1993). Hence, whereas disseminating information is one step, public learning to ensure a shared understanding of that information is an important second.

Explain discretionary elements in analyses. Whereas the preceding step is almost routine, and is often legally required, the second is more resource-intensive, less often mandated, and more often neglected. If parties are to believe that resource managers are making a good faith effort to explain the basis for their decisions, then workshops, educational presentations, and other mechanisms for two-way communication and disclosing discretionary methodological choices would seem to be essential. The *East Central Florida* water supply case describes the extensive efforts made by the regional water authorities to cultivate a shared understanding of the issues among the decisionmakers, the public, and service providers in anticipation of decisions on allocation and the development of new water supply alternatives.

Share technical expertise. If public decisions are to be legitimate and grounded in the best science, not only information but technical expertise must be accessible to ensure competency. This sharing should occur between government agencies and the public as well as among the agencies themselves. Making technical expertise available may be as simple as question-and-answer sessions with experts or as labor-intensive as tutorial sessions to interested parties who may be unfamiliar with a particular specialized line of inquiry. The *Fenholloway* case provides an example of an effort to ensure competency among the participants, with the Process Technology Work Group agreeing to "discuss the transparency of the report...including a more detailed discussion of what was feasible and on what time line, and a question of how much work and money," and environmental representatives gaining shared access to otherwise unavailable technical expertise.

Pursue joint fact finding. Joint fact finding has been utilized in many mediated negotiations of science-intensive disputes. "Stakeholders with differing viewpoints and interests work together to develop data and information, analyze facts and forecasts, develop common assumptions and informed opinion, and, finally, use the information they have developed to reach decisions together" (Erhmann and Stinson 1999). Discretionary methodological decisions are deliberated and determined by consensus. If agreement cannot be reached on appropriate assumptions, such as those used in forecasting models, analyses can substitute a range of acceptable values. Rather than bickering over the precise figure to assume, the discussion can entertain a range of probable projections, bounding a portion of the methodological uncertainty and narrowing cause for disagreement (Ozawa 1991). The early days of the *Apalachicola* conflict used joint fact finding to develop a common hydrological model acceptable to all states, although joint fact finding broke down later in the process when one side introduced new assumptions unacceptable to the other.

The products from joint fact finding are truly socially constructed by the parties and gain legitimacy as a matter of course. Involving non-experts not only enhances the technical competency of these participants, it adds importantly what Andrews calls "civil legitimacy" to the work (Andrews 2002). Knowing that stakeholding parties are involved in discretionary decisions along the course of the investigation puts observers at ease regarding the compatibility of political values and technical judgments embedded in the work.

Acknowledge the possibility of surprise. Dissemination, open access, joint fact finding, and the ability to adequately understand the work of scientists and other technical experts facilitates deliberations among represented interest groups and decisionmakers. Without them, false expectations can arise. Elected officials, in particular, often lament that the pace of knowledge accumulation and the role of surprise in scientific undertakings are quite at odds with their electoral demands. Concurrently, the temptation to strategically employ scientific work, with all its ambiguity, is often irresistible. As a result, elected officials sometimes find themselves locked into a policy position despite new contrary scientific findings because an important constituency has evolved around that position. Public statements by resource managers and elected officials, and indeed by all parties, about the routine discovery of new information, surprise, and new understandings may alleviate the pressure to be held captive to old ideas and discredited scientific views. In fact, a statement early in the process, a sort of ground rule that explains the levels of uncertainty and expectations for additional information to emerge during a given process, would be one way to avoid embarrassment. No strategy is foolproof, but early acknowledgement of possible surprises or deviations from assumptions would better protect the credibility of many participants.

Clarify scientific discrepancies publicly. When analyses or reports point to opposing policy prescriptions (otherwise known as advocacy or adversarial use of science), resource managers or elected decisionmakers ought to issue public statements clarifying the basis for the discrepancies. If they are left unexplained, the public and other stakeholders are often confused. Confusion can discredit decisionmakers or, more seriously, lead to a rejection of all scientific work and a retreat to other bases for decisionmaking. If public decisions are to be informed with the best knowledge available and the legitimacy of institutions maintained, the public must be reminded that scientists and other experts can legitimately disagree. Choosing the analysis more appropriate to the situation is a matter of judgment and values, one that ought to be deliberated openly.

Straightforward public recognition of both the possibility of surprise and disagreements among experts present opportunities for public learning. With such public acceptance, decisionmakers will gain flexibility to more aptly mirror the demands of an adaptive management system. Because this message goes against deeply ingrained expectations, repetitive efforts may be required before the public accepts surprises and a lack of unanimity in the science underpinning public decisions.

Limit decisions, and provide for review. The stakes in public decisions often loom larger than necessary because of the sense of permanence associated with much legislation, policy, and administrative decisions. Admittedly, substantial effort is needed for passage of a new bill or permit, or reversal of an existing one. Given the uncertainty that abounds in scientific undertakings involving large ecosystems, however, it is not simply hubris but folly to commit to a course of action that does not allow continual refinement or the possibility that a substantially different understanding of the natural system may arise. It may make sense, in certain cases, to limit the term of any decision, requiring a re-opening of the issue at specific intervals that coincide with anticipated new information. Alternative actions can be prescribed for a range of possible future scenarios. (See the discussion of "contingent agreements" in Chapter 14.) Required reassessment in the light of new information would constitute an adaptive approach. Such an approach would also allow entry points for those whose interests emerge later. Periodic reviews of decisions thus normalize new voices, new information, and new understandings. In this light, issuing 50-year or even 20-year operating permits for large facilities without explicit provisions for review at periodic intervals may signify a serious loss of opportunity to incorporate science into collective actions.

Monitor progress. Periodic review and reassessment make monitoring and data collection essential and eminently practical. Decisions can be refined as uncertainty is diminished. Scientific learning can proceed with political legitimacy. Without ongoing monitoring and data collection, the periodic review might be perceived as an attempt to renegotiate priorities rather than an opportunity to refine actions to achieve priorities already established.

Share risks. At times, requiring a renewal of a commitment to a past practice or inviting a re-opening of "settled" decisions may present undesirable costs to some parties and encounter resistance as a result. For example, in the *Fenholloway River,* the Buckeye pulp mill would incur costs of $39 million or more to relocate their effluent pipe. The mill would certainly not be complacent about the possibility of a retraction of the permit approving the pipe, even if scientists accumulated sufficient data over the intervening years to demonstrate irreparable ecological harm. Industries, firms, private individuals, and municipalities are sensitive to capital investment costs and rely on a certain degree of stability and predictability in decisions.

However, we may be saving dollars at the loss of far more valuable resources. Risk sharing, an arrangement that would distribute the costs of decisions and changes in those decisions, would enable decisionmakers to demonstrate responsiveness to the groups most directly affected. The *Suwannee* River Initiative is an example. Farmers were given technical assistance and economic subsidies for implementing best management practices (BMPs) to improve water quality in nearby streams. However, what the resource managers identified at that time as BMPs may ultimately prove insufficient for addressing water quality issues. If that were the case, the farmers would need to change their practices again, and would presumably incur additional costs as a result. This arrangement of technical and financial assistance and the implied continuation of such assistance in the future effectively divide the cost of the initial change and the risk of further modifications between the public agency and the farmers. Although what induced the farmers to cooperate was undoubtedly the combination of realization of the unattractive alternatives to cooperation (the legal framework, its current interpretation, and the empirical data linking water pollution to farming) and the technical and economic assistance offered, the arrangement nonetheless is an example of how risk can be shared.

Further Guidelines for Adaptive Governance

Adaptive governance requires more than technique. It requires new attitudes toward interacting with one another, and institutions restructured to support such interactions. Four areas warrant consideration.

Develop trust. Develop conditions and structures that build social trust so that parties may work together despite high levels of uncertainty in the scientific knowledge that underpins decisions. Kasperson et al. (1992) have suggested that social trust is a critical element in the siting of hazardous facilities and other contentious public decisions. They identified four constituents of social trust: commitment, predictability, competency, and caring. They contend that attending to these elements will help parties make decisions in the face of uncertainty.

In the context of public decisions, commitment and caring are closely related cousins and can be manifested as a shared willingness to recognize interdependency given costly alternatives to cooperation. This is similar to the concept of what Fisher et al. call "best alternative to a negotiated agreement," or BATNA. Participants in the *East Central Florida* Regional Water Initiative reportedly shorthanded their common BATNAs by simple reference to the Tampa Bay "water wars." Their "commitment" to the Initiative was sealed by their realization that alternatives to negotiation for each of them would be costly and not likely to generate a stable water supply. The "caring" that parties felt for one another emerged not from altruism, generosity, or love, but from a level-headed calculation of their interdependency.

Procedural regularity can be used to create a sense of predictability. In the *Fenholloway* River Initiative, the group established a process for sharing information, work groups with diverse representation, and mandatory attendance at steering committee meetings for all Initiative members. The more explicit the procedures, the greater assurance for all parties that they would have channels for influencing decisions and actions.

Finally, competency of all parties, attained through such methods as discussed earlier, ensures that the agreements are technically sound and are not likely to require revision due to misunderstandings of the scientific and technical components. Systematic attention to these constituents of social trust is important if parties are to work productively on science-intensive issues.

Focus on knowledge that matters. Embrace the limitations of science in order to get to the issues that matter. Science can be a weapon or a tool. Getting parties to use science as a tool to build a collective future rather than to undermine each other's wants requires a fundamental reorientation of public decisionmaking and the social relationships among the parties. Although the techniques cited above can be helpful, real success will come only with fundamental change in the way we talk, of the sort suggested by Susskind and Forester in this volume. We need to ensure procedural regularity and authenticity of process, acknowledge interdependencies, cultivate mutual recognition and respect, and more. It is not a simple endeavor. A close and critical examination of relevant scientific work can move us toward agreement on issues that matter most to us. As Andrews (2002) found, "In the exploration of uncertainty, the essence of the debate became much clearer."

Science and information are important and can help to clarify the real motivations behind public action. However, our focus on the use of science and information in this chapter should not blind us to the fact that sometimes the science matters very little to stakeholders. It is quite striking that the scientific basis for actions appeared to pass without question in three of the eight Florida cases. In the *Suwannee* River Partnership, for example, the farmers seemed to put up little resistance to changing their behavior in order to improve water quality. Whether the farmers were persuaded by the studies that linked water pollution to farming practices or they were simply not prepared to battle the state is unclear. In the *Ocklawaha* case, the feasibility or cost of restoring the river was not challenged; the dispute was simply whether such an outcome was preferable to the status quo. Apparently, the reservoir supporters did not want to lose the recreational value of the dam, regardless of how certain or inexpensive restoration of the river system might be. Finally, in Phase I of the *East Central Florida* Regional Water Supply Initiative, participants did not contest the District's analysis. Again, the lesson from these cases is not that science is unimportant, but that the issues that motivate stakeholders are in their essence not scientific but political. Stakeholders care who bears what costs, who reaps what benefits, and who gets to decide.

What more is needed? Dealing with these issues in the context of elite decisionmaking will only get us partway toward a system of adaptive governance. Who is included in the discussions must remain an open question, to be answered each time a conflict arises. Our political history informs us of the growing, not diminishing, diversity of voices in our polity. Demands for shared governance are growing louder, not softer. Any system we develop for folding science and other sorts of information into decisionmaking must be able to make space for heretofore silent voices and multiple ways of knowing.

Incorporate different sources of knowledge. A growing body of evidence suggests that expanded participation diversifies the types of knowledge that are considered, improving decisions, especially with respect to effective implementation and sustainability (Doak 1998; Fischer 2000). An adaptive governance system must be open to additional voices as groups previously unheard gain political strength. Our present institutions narrowly construe what sorts of information are relevant to public decisionmaking. Water quality regulations cite specific standards to be met, focus on particular species rather than the habitat as a whole, elevate some indicators and minimize the importance of others. If authentic participation is to be achieved, we must enhance the ability of our institutions and decisionmaking processes to consider diverse perspectives and multiple ways of knowing (Marshall and Ozawa 2004).

Western education has created a hierarchy of knowledge. Certain ways of knowing are privileged over others, perhaps for good reason, perhaps not. Native American scholars have pointed out the substantial gaps in evidence for many theories widely accepted among western scientists, such as the Bering Strait theory as a possible explanation for the American Indians' occupancy of the Western Hemisphere. These scholars argue that the oral history of indigenous cultures provides a way of understanding the world that is as valid as the unproven theories that are now so commonly accepted by western scientists that they pass and are repeated without critical examination (Deloria 1997). These claims are reinforced indirectly by scholars of social studies of science who point out that the peer review system often serves as a mechanism for verifying shared views, rather than establishing truth (Jasanoff 1991). Traditional ecological knowledge refers "to the cumulative body of information and insights about the natural world gained by local resource users or indigenous peoples, which is passed down through generations in an oral tradition" (Bisbal 2002, *1955*). What method of validation is more accurate? In the face of uncertainties, such judgment is difficult to pass. At minimum, all views ought to be heard. The more voices we listen to, the greater the likelihood that we will not miss something important.

Another closely related category of "non-scientific" knowledge is "local knowledge." Local knowledge, or "ordinary knowledge," is information gained by resource users through their experience with that resource over time. It has been defined as "knowledge that does not owe its origin, testing, degree of verification, truth, status, or currency to distinctive...professional techniques, but rather to common sense, casual empiricism, or thoughtful speculation and analysis" (Lindblom and Cohen 1979). Fischer has argued that local knowledge has "the potential to provide new knowledge...that is inaccessible to more abstract empirical models" (Fischer 2000). Although perhaps impressionistic, the holistic nature of local knowledge is one of its strengths.

Furthermore, while anecdotal information is often discounted derisively as "unscientific," observations aggregated can add up over time to a formidable collection that may eventually become important information for creating new hypotheses that may then be tested by conventional scientific methods. Moreover, what we call anecdotal can in fact represent a time series of data points, albeit informally collected. Favorite family fishing holes, for example, are probably good indications of plentiful fish in particular spots over several years. Anecdotes, especially based on observations about physical phenomena however obtained, should be welcomed and catalogued to be reconciled, if possible, with other existing data sets or analyses.

In addition to embracing different sources and types of knowledge, communication technique must also be thoughtfully considered and executed. Computer presentations with colorful hydrographs and maps are one way to convey certain kinds of data. Field trips to project sites convey other kinds of information. Each approach can be effective, depending on the type of information conveyed and who the receivers are. Similarly, whereas graphs and charts can condense large amounts of data quickly, sometimes stories are far richer. Processes that aim to share knowledge—scientific, technical and otherwise—must be designed to accommodate not only other ways of knowing, but also other ways of telling.

Resolve tensions between rights and environmental responsibilities. The final concern is the need to resolve what might be called an ethical tension between "promise keeping" and ecological protection (Beatley 1994). Our existing system of governance relies on legislation, administrative rules, and plans intended to create a stable context for individual decisions about private investments. In other words, individuals can regard such policies as implicit promises about future roles, rights, and responsibilities. Our developing state of knowledge about ecosystems, however, often discredits methods previously deemed appropriate for environmental protection. Recall the situation at an early stage in the *Fenholloway* River case.

Local residents and environmentalists were enraged that the Buckeye plant was allowed to continue to discharge waters they believed were contaminated with cancer-causing dioxins. They referenced a former Florida Department of Environmental Protection (FDEP) scientist who reported "massive groundwater contamination." The state scientists and administrators, however, believed the plant was entitled to continue to operate because the effluent contamination levels did not appear to exceed standards specified in the regulations. Whether or not the Buckeye plant effluents were indeed responsible for contaminating the groundwater to unacceptable levels of dioxin was not relevant to the state's position. The state focused on legal rights and liabilities, rather than conditions of the resource, and placed its obligation to the Buckeye plant owners who had abided in good faith with existing effluent regulations above a concern about current or future resource quality. This suggests a need to create conditions in which ethical conflicts can be aired and resolved.

Conclusion

The best science (and other types of knowledge) is that science whose meaning is agreed upon by the participants in a decisionmaking process. Far from making a claim for relativism, I am advocating a frank and pragmatic response to an intellectual consensus on the social nature of science that spans decades. Current decisionmaking processes and institutions lag behind our academic recognition that knowledge and meaning are socially constructed, and cause undue harm to the environment and social relations as a result.

In this chapter, I have suggested several deficiencies in conventional approaches to dealing with science and information generally in decisionmaking. I have presented arguments supporting both familiar and novel techniques for addressing these challenges, ranging from information sharing and joint fact finding to courageous public statements by leaders about the evolving nature of scientific knowledge and its consequences for public decisions. These techniques are aimed at serious deficiencies, and represent an important starting point for altering the dynamics of public decisionmaking. As discrete procedures, however, they have limited effectiveness. As the *Apalachicola* case demonstrated with its on-again-off-again joint fact finding, intermittent uses of isolated techniques do not lay a stable foundation for productive social relationships.

Broader, more philosophical changes in our attitudes toward one another as well as our treatment of scientific knowledge and information are needed. Adaptive governance calls for a system that neither elevates nor denigrates scientific work. It would allow parties to wrestle with their political differences unencumbered by arguments about scientific soundness or technical feasibility, in a way that engenders social trust among parties so as to enable them to work together despite uncertainty, with the flexibility to embrace diverse forms of knowledge generation and transmittal, acknowledging the ethical dimensions of public decisions. These Florida water conflicts indicate that we have some experience with promising techniques, but the journey toward a functioning system of adaptive governance has only begun.

CHAPTER 19

Linking Science and Public Learning

An Advocacy Coalition Perspective

Paul Sabatier

ADAPTIVE GOVERNANCE IS a complex normative framework for the proper governance of natural resources. In this book, it is used primarily to organize commentaries on the case studies and suggest ways of improving governance. As a positive (empirical) theory, however, it is relatively underdeveloped, although the case studies tend to reflect the well developed institutional rational choice perspective of Elinor Ostrom (1999; Ostrom et al. 1994). There is a lot of attention to institutional variables, but comparatively little to the belief systems of those involved, relationships of trust and distrust, and the range of scientists involved in technical advisory committees. Within this tradition, scientific learning about the world is relatively easy, so long as the institutional variables provide actors with the appropriate incentives and necessary resources. Individuals are assumed to seek their self-interest in a boundedly rational fashion. While their ability to process information is limited, there are no systematic biases in perception.

In this chapter I outline an alternative positive theory that views scientific learning by policy adversaries as much more problematic. I describe the basic tenets of the Advocacy Coalition Framework (AC), and present a set of conditions for enhancing the probability that scientific learning across coalitions will occur as summarized in Sabatier and Zafonte (2001). I then relate these conditions to some of the case studies. The discussion deals primarily with the challenge of scientific learning, the design of decision processes, and the critical link between scientific and public learning.

An Advocacy Coalition Framework of the Policy Process

The AC Framework views policy change over time as primarily the result of competition among advocacy coalitions within a policy subsystem, such as Florida water supply policy (Sabatier and Jenkins-Smith 1993, 1999). An advocacy coalition consists of interest group leaders, legislators, agency officials, researchers, and even journalists who share a set of basic beliefs (policy goals plus critical perceptions of causal relationships) and engage in some degree of coordinated behavior in an effort to make governmental policy more consistent with those beliefs. Conflict among coalitions is often mediated by "policy brokers"—actors more concerned with fashioning an acceptable compromise than with achieving specific policy goals. While the framework focuses on competition among coalitions within the subsystem, changes external to the subsystem (such as Presidential elections or fluctuations in socioeconomic conditions) and stable system parameters (such as constitutional rules) play an important role in major policy change.

The belief systems of advocacy coalitions are assumed to be hierarchically organized. At the highest or broadest level, the "deep core" of a coalition's belief system consists of fundamental normative beliefs, such as the familiar Left–Right scale, that operate across virtually all policy domains. Within any given policy subsystem, however, it is the "policy core" beliefs and the "secondary aspects" that are most critical. The former consist of basic positions, some of them purely normative (e.g., the relative importance of different values, such as environmental protection vs. economic development), while others are a mixture of normative and empirical (e.g., the proper scope of governmental vs. market authority for realizing those values), that operate across most or all of the policy subsystem. These policy core positions are very resistant to change, are only intermittently the subject of policy debate, and are usually changed as a result of perturbations external to the subsystem, although long-term "enlightenment" may also play a role (Weiss 1977). Science plays a much more important role in the secondary aspects of a coalition's belief system, as these involve disputes over the seriousness of a problem or the relative importance of various causal factors in different locales, the evaluation of various programs and institutions, and specific policy preferences.

The AC model of the individual draws heavily from social psychology. All individuals, including scientists, are assumed to perceive the world through a set of preexisting beliefs which bias the individual to see some things and ignore others (Lord et al. 1979). Members of a coalition will readily accept new evidence consistent with their views and seek to discount information which conflicts with their perception of the seriousness of a problem, the relative importance of various factors affecting it, or the costs and benefits of different alternatives. At the least, this results in "a dialogue of the deaf" in which members of different coalitions talk past each other; it can get much worse. Since members from different coalitions interpret the same evidence in very different ways (Barke and Jenkins-Smith 1993; Sabatier and Zafonte 2002), this leads members to question the motives and reasonableness of opponents. Given the fundamental proposition of prospect theory that people value losses more than gains (Quattrone and Tversky 1988), they are likely to remember defeats more than victories and thus to view opponents as more powerful than they actually are. The end result is a "devil shift" in which coalition members view opponents as more nefarious and more powerful than they really are (Sabatier et al. 1987). The devil shift exacerbates conflict between coalitions and pressures for solidarity within coalitions. Given that policy core beliefs in such a situation are very unlikely to change, the composition of coalitions is hypothesized to be stable over periods of a decade or more.

The AC Framework explicitly rejects the assumption that most bureaucrats and researchers involved in a policy area will be neutral. Some may well have no strong policy preferences, at least initially. But the framework contends that, as conflict between coalitions increases and as the interrelationships among sets of beliefs become clearer over time, initially loose groups with amorphous beliefs will coalesce into distinct coalitions with coherent belief systems. In the process, most neutral actors, particularly university scientists, will drop out. The AC Framework thus contends that, in well-developed subsystems, most agency officials and researchers who are active will be members of specific coalitions that share a set of policy core beliefs and act in concert to some degree (Sabatier and Zafonte 2001; Weible and Sabatier 2005).

Policy-oriented Learning

Policy-oriented learning involves relatively enduring alterations of thought or behavioral intentions that result from experience and/or the assessment of new information involving the precepts of belief systems (Heclo 1974). Learning is one of two processes of belief change in coalitions (the other being turnover of personnel). Learning *within* coalitions is relatively easy, as members share the same value premises and are looking for the most effective means to those ends. Learning *across* coalitions is much more difficult, as people from each coalition have different values and distrust each other. Nevertheless, the original versions of AC posited three conditions that facilitate such learning (Sabatier and Jenkins-Smith 1993).

Issues on which there is an intermediate level of conflict. Issues have to be important enough to generate sufficient research, usually by members of several coalitions as well as neutrals. On the other hand, issues involving conflict between the core beliefs of different coalitions, such as the rights of nonhuman beings or the ability of humans to improve on nature, generate more heat than light. Learning across coalitions is thus most likely on issues involving important secondary aspects of the relevant belief systems, particularly on aspects that have not previously received much attention.

Issues involving primarily the natural sciences. Cross-coalition learning is generally easier in the natural than in the social and behavioral sciences, because the theories and accepted methods are better established and the objects of study are not themselves actors in the policy debate. For example, the AC would expect more learning to go on regarding hydrological uncertainties than regarding economic estimates of the benefits from habitat restoration.

A forum that is (a) prestigious enough to force professionals from different coalitions to participate, and (b) dominated by scientific norms. The latter assures a general consensus on the appropriate rules of evidence and a minimum of ad hominem attacks, as well as some attention to underlying assumptions. Possibilities for professional forums include expert committees appointed by the National Research Council or a professional association, studies by organizations with a strong reputation for neutral competence (e.g., the Congressional Budget Office), or, in some cases, interagency technical advisory committees.

Several years ago, Sabatier and Zafonte (2001) developed the following set of conditions for a successful professional forum, in which experts from the competing coalitions come to consensus on technical or policy issues placed before it.[1] In general, such forums are most useful when a fair amount of scientific and technical information on various aspects of the topic exists, but the conclusions of various studies vary and the validity of much of the evidence is questioned by members of opposing coalitions. The arguments below assume that the essential task of a professional forum is to convince coalition experts that professional norms require the alteration of one or more beliefs important to a coalition.

Context of policy stalemate. A forum will be successful only in a context of policy stalemate or "hurtful stalemate" (Zartman 1991), when each coalition views a continuation of the status quo as unacceptable. Any party that can accept the status quo will be less willing to compromise and may not even participate.

Balanced composition. For scientists with quite different points of view to come to consensus and for that consensus to be accepted by the major coalitions, the technical advisory committee should include both (a) scientists clearly associated with each of the major coalitions and (b) neutral scientists. The chair should come from the latter.

The AC Framework assumes that the various coalitions involved in important policy disputes have scientists (or technical experts) whom they trust, presumably because they share most of the coalition's policy core beliefs and can help the coalition understand the scientific aspects of such disputes. Many interest group leaders, legislative personnel, and agency officials who constitute the leadership core of coalitions are intuitively aware of the role of values in determining the agenda of technical advisory committees and how uncertainty is treated. Thus they are unlikely to accept the recommendations of any technical advisory committee on which their point of view has not been argued by someone whom they trust. And, if they do not trust the committee's report—particularly in relatively decentralized political systems like the United States—they almost always can find some decisionmaking venue in which they can circumvent or obstruct the committee's recommendations. In short, selecting a committee composed entirely of "neutral" scientists—which often seems to be the strategy of the National Academy of Science (Boffey 1975)—may facilitate short-term consensus but will probably fail to affect policy in the long run.

On the other hand, the technical advisory committee needs to be chaired by a neutral (and probably include a few other neutrals) whose task is to impose professional norms regarding acceptable evidence, methodologies, etc. on the debate, and to indicate to coalition experts when a professional consensus is beginning to emerge. Nonprofessionals must be excluded from these deliberations in order that scientific norms can prevail in weighing the evidence and the assumptions behind different points of view.

Consensus decision rule. A corollary of the composition guideline is that the professional forum makes decisions by consensus. Any dissenting member is likely to belong to a coalition that has the resources to have the committee's decision overturned or severely compromised in an alternative venue.

Neutral funding and sponsorship. Funding must come from an institution which is not perceived as being controlled by a single coalition. This will usually require funding either by a neutral foundation or by multiple agencies that represent the various coalitions. Policy participants are unlikely to trust a committee funded and thus potentially controlled by an opponent. Since many agencies are perceived as being part of a specific coalition (Sabatier and Zafonte 2002), single-agency technical advisory committees will lack legitimacy.

Adequate duration. A forum should meet at least a half-dozen times over a year or so. It takes time for scientists from different coalitions to analyze their hidden assumptions, to critically evaluate the evidence, and to begin to trust each other. One-shot committees will probably not work.

The AC Perspective on Florida Water Conflicts

Scholz and Stiftel distinguish between "scientific learning" and "public learning." The former consists largely of scientists achieving a better understanding of natural processes, while the latter involves a better understanding of both natural and human systems on the part of not only social scientists but also other policy elites, the attentive public (including policy constituencies, such as farmers for nonpoint source runoff), and the general public. The Advocacy Coalition Framework deals largely with scientific learning. It does not say a lot about public learning, except to assume that the transmission of ideas from scientists to other policy elites and to the attentive public occurs largely *within* advocacy coalitions among people who trust each other.

To probe the question of cross-coalition learning, we will examine two case studies from the AC perspective to see whether they involve, first, a policy stalemate and, second, a professional forum in which (a) all stakeholders were represented by an expert whom they trusted, (b) there was a consensual decision rule, (c) funding came from more than one coalition, and (d) the process lasted at least a year.

Aquifer Storage and Recovery

This case involved legislation that sought to relax regulations that prohibited waste water from being injected into underground aquifers for later recovery unless it met human health standards for coliform bacteria. *Aquifer Storage and Recovery* (ASR) was a major component of the Everglades restoration plan. It involved major technical uncertainties regarding contaminants like arsenic and uranium that might increase during storage in the aquifer, the possibility that bacteria might spread to surrounding aquifers, and hydrological consequences. Despite the commissioning of scientific panels, these uncertainties were never resolved to the satisfaction of all major stakeholders and thus the bill never received legislative approval.

Of the eight case studies, this is the one whose author, Eberhard Roeder, came closest to conducting an AC-style analysis. He identified three coalitions:

- a water supply coalition composed of the South Florida water districts and water utilities, the state Department of Environmental Protection, and Governor Bush, which sought to relax the rules in order to increase the water supply for water-short urban areas in the southern part of the state;

- a preservationist coalition composed of environmental groups who were concerned about a wide array of harmful environmental effects in the Everglades and elsewhere; and
- a conservationist coalition composed of the EPA and some environmental groups, which was primarily concerned with using the aquifer for seasonal storage, but also concerned about the impacts of recovery on human health.

These coalitions appear to have been reasonably stable for several years.

To help resolve some of the scientific uncertainties across coalitions, a professional forum on the Everglades (CROGEE) was formed by the National Research Council as part of the multi-agency Comprehensive Everglades Restoration Plan (CERP). Its composition followed the National Research Council tradition of involving primarily independent scientists from relevant academic fields. CERP, which involved multiple stakeholders in a deliberative process, supported ASR as a means of Everglades restoration. Nevertheless, the CROGEE report emphasized the scientific uncertainties and thus provided ammunition for opponents. In addition, the legislature tended to engage in "advocacy science." In the end, no legislative consensus emerged.

The reasons are not altogether clear, but the AC Framework perspective suggests several likely reasons. There probably was no "hurting stalemate" for the preservationist and conservationist coalitions. While CROGEE met AC guidelines on having neutral scientists, it did not appear to be closely linked to the opposing sides in the legislative arena. The emphasis on technical uncertainty in the CROGEE report—particularly regarding hydrological impacts—did not convince advocates to seek a more modest agreement to set an agenda for resolving the uncertainties. The legislative failure to resolve the dispute appears to set the stage for further deliberation within CERP, where the presence of diverse interests may provide a better basis for resolving the scientific uncertainties in a manner more amenable to public learning across coalition boundaries.

Fenholloway River Pulp Mill Case

In the mid-1970s the Buckeye pulp mill on the Fenholloway River applied for a permit to meet new Class III water quality standards by piping its effluent discharge along the river to a site closer to the estuary. The permit was approved by the Florida Department of Environmental Protection (FDEP) in July 1997, but rejected by EPA Region V seven months later because of potential impacts on local drinking water wells and on the aquaculture industry in the estuary. On average, the mill provides about 85 percent of the flow of the river. EPA and environmental groups wanted Buckeye to revamp its production process rather than move the outfall. Buckeye argued that it had already spent $100 million in technology improvements and wanted to rely on the more cost-efficient option. Each side had its supporting studies. There followed three years of conflict and devil shift.

When George Bush was elected President in November 2000, this represented what the AC Framework terms "an exogenous shock" to the Florida water policy subsystem. Local environmentalists worried that EPA would respond by relaxing its demands on the paper mill. In short, they now faced a hurtful stalemate. This gave them the incentive to take the initiative in establishing a collaborative process among the stakeholders in December 2001.

The collaboration included a steering committee and four technical advisory committees, each composed of representatives from the four major stakeholders: environmental groups (Clean Water Network/NRDC), Buckeye, EPA, and FDEP. After two years the collaborative process substantially improved the sharing of information among participants. Particularly important was Buckeye's decision to release confidential information regarding its process technology to an engineer respected by all the stakeholders. However, the confidentiality requirement prohibited the engineer from fully informing the environmentalists about the negotiations, which strained the trust between them. In the end, no agreement was reached because the group was not persuaded by the analyses despite the participation of a technician they trusted.

Applying the AC Framework's criteria to this situation, the collaborative effort appears to have been initiated by a hurting stalemate. The membership was fairly broad, but still did not appear to include any local actors. The main technical committee was led by a respected neutral. I assume the collaborative operated by a consensus decision rule, and it certainly met the duration guideline. The case study is silent on funding, but presumably at least Buckeye, FDEP, and EPA were contributors. In sum, the AC Framework suggests that many of the critical ingredients were in place. The failure to strike an agreement underscores the difficulty involved in translating scientific consensus into a public learning process capable of legitimating a compromise solution.

Implications of the AC Framework for Adaptive Governance

Adaptive governance contends that second-order resource management problems—those arising from externalities in attempting to solve first-order problems—are generally more difficult to solve. In Florida, for example, the focus initially was on building water supply systems and draining wetlands for agricultural and development purposes, both of which produced water quality and habitat management externalities. The Advocacy Coalition Framework offers at least two reasons why such second-order problems are more difficult to solve.

First, entirely new sets of actors must interact and learn each others' constraints. In effect, Florida went from a water supply subsystem in which farmers and cities fought over supplies to a set of interacting subsystems: water supply, water quality, and habitat and wildlife management (Zafonte and Sabatier 1998). Each is composed of different sets of actors, laws, technical constraints, etc. When the relevant world becomes more complex, negotiating management agreements becomes more difficult.

Second, second-order problems are more likely to involve coalitions with very different belief systems. For example, the water supply subsystem tended to share pro-development values, but disagreed over whether cities or agriculture should benefit. Efforts to placate both coalitions exacerbated environmental problems, e.g., in the Everglades. Now, with interacting subsystems, an entirely new environmental coalition with considerable legal authority and political resources has entered the water supply subsystem, particularly on issues where it intersects with either water quality or habitat management. In fact, there is almost certainly now a South Florida water subsystem composed of the relevant actors from the water supply, water quality, and habitat management subsystems. The new subsystem has increased knowledge transfer about scientific, legal, and political issues and has probably led to some learning across coalitions. This is reflected in an ambitious, but not yet implemented, comprehensive agreement on the Everglades.

Third, given the difficulties in resolving second-order problems illustrated in both case studies discussed above, the AC Framework cautions against undue optimism about the possibilities of achieving agreement through the techniques advocated in the literature on Alternative Dispute Resolution (ADR). Recent versions of the Advocacy Coalition Framework share a common perspective with ADR on the design of professional forums and policy negotiating committees (Sabatier et al. 2005). But one must remember that the ADR literature is dominated by practitioners writing qualitative case studies (Leach and Pelkey 2001; Leach 2002), and has not been subjected to rigorous empirical testing. Some of the key propositions, such as the importance of neutral facilitators or mediators, have been confirmed by research involving bivariate correlations (Consensus Building Institute 1999) but questioned by more sophisticated multivariate analyses (Leach and Sabatier 2003).[2] Studies of 76 watershed partnerships in California and Washington have shown that how participants rate partnership success is influenced by perceptions of trust within the partnership more than by actual on-the-ground restoration projects (Leach and Sabatier 2005). This evidence of a "halo effect" supports Coglianesi's (2003) contention that "feel good" interpersonal relationships within a partnership may not necessarily translate into improvement in environmental or socioeconomic conditions. In sum, while the propositions supporting collaborative conflict resolution institutions may be intuitively appealing, the scientific analysis of those propositions is still in its infancy.

Notes

1. Many of these guidelines for successful professional forums are shared by the literature on Alternative Dispute Resolution (Bingham 1986); Carpenter and Kennedy 1988; Susskind et al. 1999). This is a case of "parallel discovery," as I had no knowledge of this literature when Sabatier and Zafonte (2001) was being written.

2. One possible explanation for the difference in findings is that the WPP study involved a random sample of watershed partnerships in two states while the CBI study was restricted to a subset of very complex and controversial cases.

CHAPTER 20

Restructuring State Institutions

The Limits of Adaptive Leadership

Paul J. Quirk

IN THIS CHAPTER, I look behind the events of these thought-provoking case studies to consider some of the fundamental institutional arrangements that they take largely for granted. One can only applaud the effort to understand and improve the methods of adaptive governance—conceived as a collaborative approach to management and leadership that takes the basic state-level administrative institutions for water governance as given. Even more, one must admire the efforts on the part of many state, regional, and local officials to practice adaptive governance—often in remarkably difficult circumstances. That they scored some successes is testimony to their dedication. Nevertheless, state leaders have overlooked the limitations of ad hoc collaborative processes.

In my view, Florida's water governance system has come to depend on extraordinary feats of negotiation. Put differently, the state has not established the necessary policy and administrative conditions for local and regional water managers to design effective decision processes—ones that can reach sufficient consensus to implement decisions and achieve policy responsiveness—for the current generation of water issues. To be reasonably successful in dealing with the current generation of water conflicts, the state will need to strengthen its institutions of representative democracy, especially its statewide administrative institutions.

Development and Performance

To begin with, Florida's water management system undoubtedly has been successful in many respects. The regional water management districts have been generally effective and often innovative in developing water resources and regulating their use. They are often held up as a model of regional coordination. For most of their history, however, water resources in Florida have been plentiful. Even then, the districts' performance was compromised by a tendency to over-permit—that is, to grant permits for more usage of water than was possible in the long run. This over-permitting figured prominently in the state's recent encounters with scarcity. It is hardly remarkable that regional water management districts looked successful when they were granting generous usage permits in a period of plentiful water resources.

Moreover, as the case studies make abundantly clear, the Florida system has had great difficulty with the larger-scale, more complex water conflicts of recent years, as scarcity and environmental vulnerability have become impossible to overlook. Decisionmakers often have been unable to reach agreement at all, much less to reach efficient and workable solutions. Among the eight case studies, several are stories of frustration and failure (*Tampa Bay* "water wars," *Ocklawaha, Apalachicola*[1]); some are partial successes or have outcomes yet to be resolved (the *East Central Florida* Water Supply Planning Initiative, the *Suwannee* River Partnership, *Everglades* Restoration). Some involve issues of lesser scope (the *Fenholloway* River Evaluation Initiative). To my reading, there are no instances of clear-cut success on issues of broad scope that affect multiple agencies. At a minimum, the cases demonstrate that there are now major conflicts over water resource management that the existing institutions for policymaking appear largely incapable of dealing with.

At bottom, the difficulty of water management in Florida arises from the juxtaposition of two circumstances. First, Florida has developed a decentralized and fragmented administrative system for water management. As the recent report of a prominent advisory group (Florida Council of One Hundred 2003) pointed out, the 1972 legislation that provided the basis for subsequent development was based largely on a model statute, drafted by an academic, which would have created a statewide commission with significant powers. The legislature, however, left that provision out of the law, evidently finding the commission unnecessary or impractical. Instead, it provided for the five regional water management districts, with a variety of planning and coordinating responsibilities but limited powers to dictate policy to local governments. The districts' jurisdictions correspond to major river basins, but not, therefore, to the aquifers that also supply much of the state's water. Furthermore, the law imposed a strong preference for local use of water supplies, discouraging (though not prohibiting) transportation of water across county lines.

Second, the state has experienced enormous population growth, along with expanding industrial and agricultural uses of water. The population of Florida has doubled since 1972. Moreover, the growth has occurred primarily in the southern parts of the state, where water is most scarce, creating a huge disjunction between the availability of water and the location of people wanting to use it. Industrial and agricultural development has not only intensified demand, but has also generated serious pollution of some water resources. Shortages have developed, recreation areas have been threatened by overexploitation, and environmental issues have become contentious.

In this context, crucial tasks require coordinated action among multiple local governments, regional water districts, and functional agencies. In addition to having certain common interests, for example in avoiding excess exploitation, they all have distinct distributive interests. One municipality is using water;

another wants it. One agency is draining a swamp for agricultural use; another wants it for natural habitat. Judging from the accounts of the cases, there is often little likelihood of authoritative guidance from the legislature, the governor, or a statewide agency. So the various agencies set up ad hoc collaborative processes in the hope of reaching agreement, but the circumstances are far from promising for a successful outcome.

Issue Characteristics that Inhibit Agreement

These water conflicts often have several features that militate against cooperative resolution. (For a general treatment of conditions for cooperation in public policy conflict, see Quirk 1989 and McFarland 1993).

Low-information constituencies with limited opportunity for public learning. In most water conflicts, political representatives negotiate on behalf of various mass constituencies, such as the citizens of a municipality, the farmers in a river valley, or the members of an environmental group. Such constituencies may be highly concerned about their interests in the dispute. Citizens of a municipality, for example, will become quite excited about their water bills, their access to sufficient water for swimming pools and lawn sprinklers, or their opportunity to build new houses in outlying areas. At the same time, these mass constituencies pay little attention to the details of negotiating strategy and are relatively oblivious to the constraints on negotiating partners. They are unlikely to have realistic expectations about the ability to extract concessions from opponents. Thus they induce rigidity and severely conflicting demands on the part of their negotiating agents, the political officials who represent them.

Realistically, most members of such a constituency will not participate in or pay attention to a collaborative or other learning process enough to overcome this rigidity. No more than a few individuals will take the time; the others will be occupied earning their livelihoods, spending time with their families, or even engaging in other political or community activities.

Predominant conflicting interests. In any negotiation, the opposing sides have conflicting interests (for example, labor wants higher wages, management wants lower wages); they also have common interests (both sides want to avoid a strike). If their common interests are more central, the parties will more easily reach agreement. If their conflicting interests are more prominent, they will focus their energy and strategies on the *distributive conflicts*—getting more of what is in dispute between them—and are less likely to reach agreement.

Although generalizing is hazardous, it seems that the conflicting interests are generally more prominent in water management issues than the complementary interests. Taking issues of access to water once again as a model, two communities both want more water for immediate use. Figuratively speaking, they're thirsty, with the sense of urgency that implies. They also want to avoid harming the aquifer they both draw from. Perhaps its water level is abnormally low. But the conflicting interest in more water is clearly the more vivid, immediate concern. In general, the common interests at stake in water conflicts are less immediate, longer term, even somewhat hypothetical: if we keep drawing down the water level, at some point the aquifer will become unusable; if we stand in the way of efficient development, it will impose a burden on economic growth; if we don't reach agreement soon, it will prevent industries and communities from being able to plan, and so on.

Advocates of consensus processes sometimes argue as if the conflicting interests in a dispute were, at some level, illusory. They suggest, for example, that apparent conflicts reflect largely the polarizing effect of adversarial governmental processes; or that such conflicts will recede in importance if the parties to a dispute have adequate opportunity and professional assistance to develop trust, loosen commitment to negotiating positions, and discover new alternatives. Such instances are probably fairly rare. Conflict certainly has a polarizing effect, which is one reason that consensus processes can be useful in undoing the initial polarization. But basic conflicts over water management, or public policy generally, are not often mainly figments of imagination. Communities face real choices between agricultural and industrial use, between environmental quality and low-cost supply, and so on. If there is just one glass of water, then I get to drink it, or you do. Generally, therefore, the parties to water management disputes will pursue their conflicting interests and overlook their common interests.

Asymmetric stakes in the status quo. I would argue that resolutions are harder to achieve in water conflicts because one or more parties usually sees a major advantage in the status quo. If one party is much better off than the other under the status quo, it will have little incentive to reach agreement, and the negotiation is likely to fail. In principle, of course, the weaker party could recognize its disadvantage and make up for the stronger party's rigidity by making most of the concessions. But the parties are likely to differ in their perceptions of the strategic advantage. For example, the weaker party may hope that, in the absence of agreement, a lawsuit or an act of the legislature will eventually turn the tables. Such hope will raise its expectations for a satisfactory agreement. The stronger party may fail to reckon with such possibilities.

Such asymmetries in satisfaction under the status quo appear characteristic of water conflicts. One municipality has plenty of water, another has shortages; environmentalists have secured a ban on development, and businesses or communities are trying to relax it; farmers have a permit to irrigate their fields, but environmentalists want to reduce water consumption. In the absence of an agreement, all parties keep doing what they are already doing; the advantaged party is largely satisfied, even though it would prefer a sufficiently favorable permanent solution. Such negotiations are harder to resolve.

Incommensurable equity claims. Most negotiations are not merely matters of power. Especially in a political environment, the parties also pay attention to the comparative equities. If one party's claim seems to have merit, the others will give that claim some recognition for that reason alone. Indeed, even if the dispute poses conflicting, yet familiar notions of equity, those notions will help identify a plausible range of defensible outcomes. People are accustomed to conflicts between claims of need and those of merit, for example, in access to educational resources. All sides will recognize that both sorts of claims normally receive significant weight, and will dismiss positions that ignore either notion entirely.

In some water management conflicts, the grounds for the opposing claims are fairly distinctive to the conflicts, and relatively new to the parties. In disputes over access to water, for example, opposing parties may lay claims based on current access, the location of the water resource (an aquifer that runs through one county and not the other), prior investment in developing the resource, or future or projected need, among others. Without a long history of such conflicts, the opposing sides will have difficulty comparing the claims. Each actor, and especially its mass constituency, will see its own claims as absolute. The aquifer runs under our county; therefore it's our water, period. To the extent that water conflicts involve such incommensurable equity claims, negotiations are likely to be more intractable.

Multiparty Consensual Processes

The most debilitating circumstances in the Florida water management cases, however, are the sheer complexity of the negotiations—the number of independent parties, as well as the number of issues to be resolved—along with the lack of any clear-cut, governmental, default decisionmaking process.

The problem with relying routinely on large-scale collaborative processes, premised on an expectation of consensual decision, is that it amounts to an informal constitutional requirement of something approaching unanimous consent. In a world of complex, conflicting interests, such a requirement is simply untenable as the basis for effective governing. Western political thought has recognized the fundamental advantage of majority-rule (or moderately qualified majority-rule) institutions since at least John Locke's *Second Treatise of Government* (see Buchanan and Tullock 1962; for discussion of various qualifications of majority rule, see Lijphart 1999). American political thought has recognized it since vivid lessons were learned under the Articles of Confederation—which one can think of as the *Tampa Bay* Water Wars writ large. Each of the 13 states had to consent to any important decision; as a result, government under the Articles was a failure.

The central problem of a multiparty consensus process is that, if consensus is genuinely required to support action, it gives any agency or group a tremendous incentive to hold out. Suppose that three actors must agree if a decision is to be made and their common interests served, and that two of the actors are approaching agreement. The third actor then has an opportunity to exploit the situation, for example by offering to pay somewhat less than his proportionate share of the costs. The first two actors have an investment in their impending agreement, and an opportunity to achieve joint gains, and may decide to tolerate this exploitation. Now suppose, however, that there are ten actors, and nine are approaching agreement. The tenth actor can exploit the first nine even more readily. The first nine, seeing the exploitation coming, may withhold concessions they would otherwise make. Such opportunity for exploitation nearly always obtains in multiparty unanimous consent (or nearly unanimous consent) processes. In general, such an institution will be unable to act except in extreme circumstances, when everyone is forced to relent in order to prevent calamity.

Consensus-seeking processes have extraordinary advantages, provided that they supplement, not replace ordinary governmental processes. Learning, creativity, and flexibility are induced by the opportunity to do better for all concerned; but an ordinary governmental process that will eventually lose patience and impose a decision plays a vital role in moving the consensus process along. It is not sufficient that a regular government process operates at some far remove from a local dispute—for example, that the legislature theoretically could step in and resolve it. The consensus process should be clearly tied to a specific governmental entity that is watching the dispute and is ultimately responsible to act.

We can see that well-defined governmental authority is central to the Florida cases. Sometimes it is fairly well defined, and ready to take control as soon as consensual negotiations fail. The difficulty of decisions is then noticeably reduced. In dealing with the emerging problem of nitrogen runoff from agricultural lands in the *Suwannee* River Basin, for example, we find a multiparty negotiation among the Florida Department of Agriculture, the Florida Department of Environmental Protection, and numerous farmers' cooperatives about requirements for limiting runoff. But in fact the default authority was fairly clear. The environmental agency had the legal authority to adopt regulations to control runoff, although it arguably lacked the enforcement resources and political support to impose a solution. The agriculture agency was interested in an accommodation. So it helped farmers form cooperatives to identify satisfactory practices, submit proposed agreements to FDEP, and monitor compliance. In the end, the environmental agency had the authority either to accept the proposed self-regulatory agreements or to proceed with a traditional regulatory process. If the self-regulation arrangement fails, in such a case, the fault does not lie with the institutional structure, but with the deficiencies of the political support and financial resources of the lead agency.

However, in several of these cases there were contentious multiparty negotiations with no obvious default decisionmaker. To be sure, there is the possibility of an appeal to higher authority. If negotiations flounder long enough, with dire consequences, the legislature may impose a solution, as in the *Ocklawaha* River Restoration. But such possibilities are so vague and so distant that they probably figure minimally in participants' calculations. In the *Tampa Bay* "water wars," six local governments sought unsuccessfully for two decades to reach agreement on access to water resources. The conflict was complicated by important advantages in the status quo: Pinellas County had purchased land and developed well fields in two nearby counties. The state legislature's perfunctory contribution was to create a cooperative regional water supplier in 1974. But this failed to resolve the conflicts until it was transformed, by agreement among the six local governments, into a special district in 1998.

It is instructive to consider what lesson Florida's political leadership took from this episode. The *East Central Florida* initiative, established in 2002, is an effort to avoid a similar fruitless conflict. The supposed solution has been to create a collaborative, participatory planning project involving local governments in ten counties and five regional water management districts, with an ambiguous role for FDEP. No important results are yet in. But near the starting point, the

Initiative looks like an effort to overcome the difficulties of a large, complex negotiation by creating an even larger, more complex negotiation.

The mother of all complex negotiated processes is the Everglades Restoration. This undertaking encompasses "16 county governments, 122 cities, two tribal governments, a regional water management District, numerous special water supply Districts, five regional planning councils, five major state environmental planning and regulatory agencies, and eleven federal agency managers... [I]n 1995 there were approximately 200 plans addressing the storage, treatment, distribution, and conservation of water in South Florida." (*Everglades*, Dedekorkut 2003, *230*). Much of this complexity was inherent in the massive scope of the project. But state leaders left most of the planning and coordinating to a series of huge consensus-building efforts—including a ten-year stretch of mediation in the 1980s—with minimal direction from high-level state agencies and planning commissions. Progress required intervention by the governor and the ensuing Governor's Task Force, which laid the necessary groundwork for the Comprehensive Everglades Restoration Plan.

If the complex interregional, intersectoral, and interfunctional conflicts considered in this book are the future of water governance in Florida, it is impossible to maintain that the current institutional structure is workable.

Prospects for Reform

Once created, basic institutional arrangements are hard to change. There are, however, early signs of movement toward significant reform in the institutions of water management in Florida. A business-oriented advisory group, the Florida Council of One Hundred (2003), has issued a report highly critical of the state's water management system. It calls for the creation of a state water commission with significant regulatory and management powers (roughly speaking, the provision omitted from the 1972 model bill), and for modification of the policy discouraging transportation of water across county boundaries. Governor Jeb Bush cautiously said only that the state needs to "think about" these proposals. To venture a prediction, it seems to me inevitable that Florida will adopt these reforms, or something like them, before too many more years elapse. The basic facts that best explain the inadequacy of the existing arrangements—that the state population and economy have doubled since 1972, and that most of the growth is in the south while most of the water is in the north—also point to the political forces that should eventually overturn those arrangements. Most of the population and most of the economic power of the state will eventually demand change.

Such a statewide commission would presumably replace the state's informal unanimous-consent institutions for complex water issues. To spell out the precise structure or authority of such a commission is beyond the scope of this essay. Clearly, however, it would not take over most of the functions of the water management districts, or simply centralize water management at the state level. The complexity of water problems and the importance of local circumstances would make such an approach unmanageable. Rather, a state water commission likely would perform several roles. It would set general policies, goals, and guidelines for local and regional water management. For example, it would set minimum standards for sustainability of use, certify technical methods of estimating sustainability, and develop research and monitoring standards to determine sustainability. It might also impose constraints on water management policies that fail to provide for reasonable amounts of economic and population growth. It would review local and regional water plans and agreements, and have the authority to amend or reject them.

In view of the extensive history of collaborative efforts in the state, a state water commission should have a mandate to encourage and support collaborative processes at the local and regional level, and to sponsor new collaborative efforts at the state level. In particular, it should establish a collaborative process to consider the fundamental conflicts between northern and southern counties over long-distance transportation of water. Crucially, it should have authority to set deadlines for consensus processes and ordinary planning decisions, and to impose decisions when local and regional institutions are unable to overcome deadlock. Finally, as a residual authority—which it should invoke only in cases of severe local or regional stalemate or refusal to comply with state guidelines—it should have authority to impose plans of its own devising.

Rather than simply supplanting collaborative processes at the local and regional level, a well designed state water commission would make those processes more effective. By providing deadlines for agreement, it would remove one of the main difficulties with existing collaborative processes. By revealing its own general policies toward the tradeoffs in water policy, and holding out the implicit threat of imposing a decision as a last resort, the commission would also give the parties to consensual processes a prominent focal point for the design of possible solutions—and most important, a broad sense of each party's relative strength in the negotiation. That effect means that the commission would alter the distribution of power among parties to water conflicts. But the changed distribution should better reflect the priorities of the state. The distribution of power under existing processes tends to reflect each party's willingness to delay agreement and hold out for a better deal. In such a context, collaborative processes would play a very useful role. They can often permit exchanges of ideas and suggestions that would never be possible in the charged rhetorical atmosphere of policymaking in a legislature or a state regulatory proceeding. But it is the relatively immediate prospect of an administrative proceeding, in the background, that makes collaborative processes work.

Finally, the commission would provide an appropriate venue for discussion of the large-scale, high-politics water management issues facing the state. How far does Florida want to expand agricultural production in the north or population in the south? How much economic development will the state sacrifice to restore the Everglades to a more natural state? Inasmuch as water cannot be made perfectly pure, or even perfectly free of harmful substances, what level of purity does it want to insist on, and what level of risk is appropriate for injecting water into aquifers? Are federal standards high enough, even as they fluctuate with the political fortunes of various groups, and especially the major political parties, in Washington? In the end, these questions should be decided by the legislature, the governor, or even the voter. But having a state water commission, with responsibility to give advice on broad matters of policy, would assist them in their deliberations. If nothing else, commission decisions that affected these issues would prompt the legislature and governor to address them.

Florida's water managers cannot wait for such reform to happen. The citizens for whom they work need water, and they need water-related problems addressed immediately. Responding to those needs is the manager's day job. Responsible, progressive managers need to divide their attention between using adaptive management and adaptive governance to secure the best outcomes possible under existing institutions, on the one hand, and promoting and preparing for fundamental reforms in Florida's water governance system on the other.

Notes

1. Because the *Apalachicola* Basin case concerns an interstate conflict and negotiations under federal auspices, it does not reflect on the performance of Florida's institutions. It is an example, however, of the type of multiparty negotiations over issues of large scope that seem to overtax them.

CHAPTER 21

Incentives and Adaptation

Lawrence S. Rothenberg

THE IDEA OF ADAPTIVE GOVERNANCE—a new generation of governance institutions capable of dealing with so-called second-order problems—raises a litany of questions. Does the American political system respond to increasing and competing demands by adapting its institutional structures efficiently and equitably? More specifically, can the bureaucratic apparatus in Florida, operating within the "adversarial legalism" of the American regulatory system (Kagan 2001) and its constitutional separation and fragmentation (Moe and Caldwell 1994), reinvent itself to meet the policy challenges presented by rising consumption of, and competition over, scarce water supplies and resources and by technological hurdles and uncertainty? Should we view events in Florida optimistically as adaptation to changing issues and stakeholders and, ideally, can we provide helpful and feasible policy recommendations?[1]

As an analyst, I can answer some of these questions better than others. I come to the water conflict debate as a student and observer of similar issues debated at the national level for comparable reasons—the desire for more cost-efficient, cooperative, and comprehensive processes that bring many stakeholders to the table and for institutions and solutions that meet their needs—and myriad attempts to address them (e.g., Glazer and Rothenberg 2001; Rothenberg 2002). Hence, I will concentrate on developing three principal points regarding the situation in Florida.

First, the commonality with general environmental policy is striking. We are observing real politics in a world where the desires of stakeholders—firms, governments, environmental groups, citizens, and consumers—increasingly clash, choices often seem maddeningly inefficient and occur at a snail's pace, and escalating costs nonetheless induce key players to search for less expensive, more efficient, solutions. Efforts to set up more flexible, responsive, and economical bureaucratic processes tend to move slowly and produce mixed results. Apparent successes at one moment often unravel with new events or when opportunities arise for actors to better themselves. There is little in the Florida case studies to indicate that a new model of agency behavior has evolved, or is likely to. Rather, events in Florida are consistent with increasing national efforts to find solutions that modify but do not qualitatively alter how choices are made, so that stakeholder costs are minimized and politicians can appear responsive to their constituents. Our Florida case studies may reveal general factors facilitating or undermining adaptive governance, and the national political experience can provide insights relevant to Florida.

Second, aspects of our case studies where adaptive governance appears to be more successful (representation, learning of various kinds, and problem responsiveness), coincide with meaningful rewards and punishments being available for enforcement. Credible commitment to rewarding cooperators and punishing defectors is a well known factor for developing long-term alliances in other contexts, and it appears to be crucial here.[2]

State solutions in Florida differ from their national counterparts in their unwillingness to consider market instruments. Whether oriented toward price or quantity, market mechanisms represent an effective way to manage many environmental issues, and adding them to the political toolbox would be a good thing. So a third issue is whether market-based solutions can function as a means of adaptive government that helps resolve water conflicts.

The Context: Conflicting Demands and Awkward Supply

The root causes of the problems confronted by Florida water interests are straightforward, and the conflicts and reconciliation efforts are comparable to those defining environmental matters in the United States. Indeed, what we have learned about American environmental policymaking can help us understand water conflicts in Florida.

Competing demands for water resources have increased, creating what Scholz and Stiftel term second-order problems. Efforts to provide a host of stakeholders with access while maintaining and enhancing environmental quality not only involve technological hurdles and scientific uncertainty, but also necessitate overcoming political fragmentation and myriad related political economic problems common to most environmental policy efforts.

The heightened pressure on Florida's water resources is most visible in demand for usage. With population growing roughly two million in the 1990s and incomes increasing, demands for water have increased and diversified, ideally cheap water from the perspective of those paying. Expanding localities fight each other, burgeoning urban areas compete with agricultural interests, citizens confront industry, recreationalists confront environmentalists, and even states fight each other. Each tries to abrogate others' property rights in favor of their own, and generally to cut the best possible deal for themselves.

Yet, consistent with the idea that Americans have become a nation of "Lite Greens" (Ladd and Bowman 1995), pressures for a quality environment have also grown, frequently among those wanting water resources for their own purposes.

The reason for this increased demand for environmental quality is straightforward: quality is essentially a normal good; therefore demand, particularly with greater sensitivity based on increased scientific understanding, rises with prosperity (Kahn and Matsusaka 1997). The nature of second-order problems can be seen not merely as the product of the complexity of environmental problems and the problems created by greater usage of natural resources, but also as a function of changes in our preferences about environmental quality. For example, we are both more aware of and less willing to tolerate despoiling water bodies through effluent dumping (*Fenholloway*) and are suspicious of technologies that might threaten aquifers (*Aquifer Storage*).

Additionally, the development of many citizen groups is more tangible evidence of greater demand. This growth of environmental interests (and the responses by those threatened) mirrors national changes (Berry 1999). As we see in numerous Florida cases, groups, some with purely local constituencies and others with national audiences (Sierra Club and Natural Resources Defense Council, and others), have helped transform policy debate and the political and administrative processes required for dispute settlement. For instance, we find Proctor and Gamble denouncing as a "pseudo-environmental cult" the local group Help Our Polluted Environment in the debate over *Fenholloway*, where group efforts garnered national attention on *60 Minutes* (an event rarely welcomed by a large corporation).

We observe an inherent and increasingly prominent tension between water resource exploitation and environmental protection. In some cases, such as a paper plant dumping effluent into the Fenholloway River, this tension is easily explained, as pollution is a byproduct of production. In others, such as the Everglades, where a wide range of actions put entire ecosystems at risk, the relationship between greater consumption and utilization of water resources and environmental pressures is more subtle and subject to scientific uncertainty and dispute. But the commonality is a tension between resource exploitation and environmental quality, with society demanding actions to ensure both.

This, in turn, makes the government's role in the twenty-first century more complicated and intrusive than for much of the last century. Much earlier government activity was designed to encourage economic growth and activity by facilitating exploitation of water resources—most notably, via large water projects under the aegis of the Bureau of Reclamation and the U.S. Army Corps of Engineers. This was true of the national government's actions regarding the environment generally: in a hopefully sustainable fashion, land was to be productively used, natural resources exploited, and water resources harnessed. Now, rather than a symbol of progress, environmentalists tend to consider such actions (and the institutions producing them) irresponsible, as in the *Ocklawaha*, where they engage in largely unsuccessful efforts to undo a project for which vested stakeholders now exist. Such cases again mirror national efforts, notably widespread initiatives to breach dams and undo efforts to take advantage of water resources (Graf 2002; Lowry 2003).

Government's more intrusive role is seen both nationally and in Florida. Nationally, the U.S. Environmental Protection Agency (EPA) has been created and expanded, and other government agencies dealing with natural resources have been transformed; in Florida, the state Department of Environmental Protection and a host of local government players have been created. Both levels have promulgated detailed regulations and programs designed to influence water and ecosystem quality.

But government efforts at improvement are notoriously frustrating. Incentives conflict in an institutionally fragmented political system. At the national level, environmental quality efforts are conspicuously hamstrung by these conditions, and we see much of the same in Florida. Indeed, in these case studies, as in environmental policy overall, we find numerous instances of interaction between national, state, and local actors within the complexity of the federal system contributing to this dysfunctional picture.

Furthermore, with water issues becoming more costly and conflicted in states such as Florida, there is increased desire to reform institutional mechanisms to reduce costs and conflicts by developing trust, cooperative decisionmaking processes, and more flexible ways of dealing with technical complexity. These efforts, too, mirror national reform and consensus-building efforts.

In short, these case studies appear to be textbook examples where increasing demands for consumption and environmental quality produce awkward attempts by government to supply both. There is a strong desire, given rising demand and inadequate supply, to see institutional structures adapt to a rapidly changing world. The issues of adaptive governance are general phenomena, not unique to water conflicts.

Core Elements

In the case studies of adaptive governance we see standard features of contemporary American environmentalism. These features represent potential obstacles to the evolution of political and deliberative structures and ultimately the resolution of second-order problems. It is difficult to summarize them all neatly, but let me highlight a few core elements.

Fragmentation

Perhaps the most notable feature of the case studies is fragmentation created by the expansion of societal and institutional stakeholders. Such fragmentation is also a defining feature of national environmental policy (Davies and Mazurek 1998). There are conflicts between national, state, and local governments; between different localities; and between environmental and business interests. We witness a host of constraining environmental rules (*Fenholloway*) that often highlight a tension between accountability and flexibility. We typically see constraining, sometimes contradictory, environmental statutes and obligations—as highlighted by the *Ocklawaha* River Restoration case where there are "two legislative mandates directing mutually exclusive outcomes" (Sloan 2003).

Furthermore, we see the ability to agree undermined by the multiple venues in which interests can influence policy by either changing or maintaining the status quo. Sometimes stakeholders appeal to state and local agencies, sometimes to the EPA, and still other times to state or local politicians. And in numerous instances the threat of adversarial legalism and quasi-judicial or judicial remedies further complicates matters. Moreover, when all else fails, groups can resort to "private politics" (Baron 2001)—rather than appeal directly to government, they pressure others, notably firms, through adverse publicity (*Fenholloway*).

Importance of the Status Quo

Throughout our case studies, changing the status quo is difficult; in numerous instances the actor whom the status quo favors fights change and contrives delays. Indeed, maintaining the status quo seems to be a key element in many instances of seeming governance failures. More generally, in a political system favoring the status quo, we need to consider how adaptive governance can advance problem responsiveness when one stakeholder prefers the current state of affairs.

Established Interests due to Investment

Investments, once made, often create interests that find alternative solutions unappealing. Thus, when corporate stakeholders invest large amounts of money in a pipeline in the *Fenholloway* case, they are loath to consider the desire of environmentalists to reduce initial pollution through other means.[3]

As the discussion of these core elements underscores, not only does the Florida policy process share many features with environmental and public policy generally, but these features may thwart institutional structures that would allow for learning, representation, and responsiveness. At a minimum, a well-designed system will need to overcome fragmentation, allegiance to the status quo, and reluctance to lose existing investment.

Sources of Success and Failure: Establishing Commitment and Credibility

Although my comments may sound pessimistic, clearly some Florida cases produce more successful solutions than others. Furthermore, sometimes the institutional structures chosen seem important, particularly as a way to foster trust and cooperation. While some factors that cannot be controlled may play a role in achieving success—relative technical complexity, the number of stakeholders, the homogeneity of their preferences, or pure randomness—it appears that some manipulable factors are at work.

How important are these other factors? Certainly, cooperative solutions are attractive, although not the only, means of adaptive governance, and social scientists have established that trust and cooperation are possible, but not assured, given repeated interactions and sufficiently modest discount rates (Fudenberg and Tirole 1992; Scholz and Lubell 1998). But many cooperative outcomes depend upon the willingness and ability of key actors to reward and punish defectors. While efforts designed to induce a consensus might be sufficient for producing learning, sustaining the long-term relationships necessary to achieve the ultimate goals is likely to prove problematic without threats and promises. Indeed, the threat of punishment or lure of reward is strongly evidenced in cases with successful outcomes:

- In *Tampa Bay* (successful in that a palatable solution is arrived at, though after much delay and cost), financial assistance, threats of legislative retribution in the form of a statutory solution, and threats of agency retribution by the Southwest Florida Water Management District when the dispute finally moved toward resolution, all seemingly played a role in creating a viable public utility.
- In *East Central Florida*, financial incentives and the threat of permit restrictions may be needed for long-term success; otherwise the situation may look more like the initial morass in Tampa Bay than a victory for rational planning.
- In *Suwannee,* there are threats of a legislative or regulatory solution, and rewards in the form of financial subsidies and technical assistance: "Trust, voluntary cooperation and cost sharing are key elements of the Partnership (Hemphill 2000). Success of this program is recognized as being largely dependent upon funding subsidies and technical assistance" (SRWMD 2003b as quoted in Dedekorkut 2003, *103*).
- For the *Everglades*, although the provision of $200 million is mentioned, it should be highlighted that the national government is picking up half of the estimated $8 billion tab for restoration as the glue in a cooperative solution.[4]

There is one case in this volume where collaboration appeared to be heading to success *without* threat or reward—*Fenholloway*. The Evaluation Initiative is depicted as an expensive means of fostering trust and a solution to a decades-old problem. The background of the case indicates that collaboration was an option of last resort for an opposition that faced a firm favored by the status quo, and had seen its court cases thrown out, attempts to rally the media prove insufficient, and national support evaporate through changes in presidential administration and at the EPA. With little negotiating leverage, Linda Young essentially sued for peace.[5] In the end, even this case ended with no agreement—the political climate continued to shift against environmental interests, and what was attainable in the Initiative was not acceptable to them. Incentives to force an agreement acceptable to both parties were not sufficient.

These observations suggest that cooperation works in a game-theoretic sense: deliberative institutional structures function better when punishments and rewards are enhanced. We must think carefully about what is driving stakeholders to a more collaborative stance—is it a trustful environment fostering a "win-win" solution by itself (i.e., where opportunities for mutual gain are realized) or is it this in conjunction with the prospect of threats and rewards? While only suggestive, it appears that carrots and sticks—provided or withheld according to clear standards—are crucial for getting stakeholders to participate in more deliberative processes that produce lasting solutions. This in time could become a normal expectation if recognized as key for consensual processes.

Addendum: Experiences at the National Level

The Common Sense Initiative is a fundamentally different system of environmental protection... government officials at all levels, environmentalists, and industry leaders will come together to create strategies that will work cleaner, cheaper, and smarter.
EPA Chief Carol Browner announcing the Common Sense Initiative (CSI), July 20, 1994[6]

National efforts similar in spirit to Florida's efforts at adaptive governance offer additional perspective. As in Florida, these examples typically involve trying to forge consensus to take advantage of participants' knowledge and understanding, enhance flexibility, and reduce transaction costs. Bill Clinton's election in 1992 initiated a movement to "reinvent government" which, applied to environmental policy, was essentially adaptive governance (National Performance Review 1993). EPA's CSI was announced with great fanfare as the centerpiece of the agency's reinvention efforts (Project XL was its other major initiative), and had many characteristics that one would link to adaptive governance. Specifically, it was designed as a means for individual industries and multiple stakeholders to deal with scientific uncertainty, reconcile competing stakeholder interests, foster creativity, and produce problem responsiveness.

But what was the reality? Although the EPA lauded the program, more dispassionate observers were less sanguine (GAO 1997; Coglianese 2000; Coglianese and Allen 2003; for a comparable analysis of Project XL, see Marcus et al. 2002). Perhaps most interesting are the two principal reasons given for program shortfalls:

- Trying to improve policy within the constraints of the current environmental legislation is a considerable obstacle. Without better designed legislation, the CSI's effectiveness is inherently limited.
- When participation is voluntary, consensus building tends not to be very effective and comes with high transaction costs. "Deliberative, sector-based efforts such as the CSI may well serve a useful purpose of generating some new ideas, making incremental changes, or providing feedback to those involved in the regular policy process, but we should not expect that consensus-building will provide the route to a fundamentally 'cleaner, cheaper, and smarter' regulatory system" (Coglianese and Allen 2003).

Adaptive Governance by Other Means: Instrument Choice

Economic growth brings abundant benefits but can also unleash a wide array of environmental problems... The challenge in addressing environmental problems lies in harnessing and channeling the power of markets, so that they both deliver continued economic growth and foster sound environmental practices.
U.S. Executive Office of the President, 2000

We finally consider how adaptive governance might be achieved by other means. Are there ways to break out of the maze of incremental changes and iterative learning and find a more efficient and, arguably, more equitable model?

One prominent way in which Florida has diverged from the national direction is the lack of attention paid to market-oriented reforms, which are dismissed as unfeasible. Nationally, and in other advanced industrial nations, such solutions are being proposed and implemented, from tradable permit markets for acid rain to market-based instruments to combat global warming. This process involves politicians and their allies assuming a leadership position, winning over initially skeptical environmental groups, and building coalitions with business interests to support more efficient environmental programs. The decision process issues focus principally on setting goals (e.g., the amount of pollution to be reduced or water consumed) and establishing the framework of a market (e.g., a market for trading).

Hence, while in Florida approaches with different degrees of cooperation and conflict have been proposed and utilized, all such solutions are bureaucratic and all regulatory instruments are to some degree command-and-control. Elsewhere, market solutions have become a widespread, if not predominant, means of dealing with technologically complicated, very expensive, environmental problems (Kirchgässner and Schneider 2003; Stavins 2003). These are not necessarily a panacea, and require careful design by those adept in such incentive compatible mechanisms, but markets do offer possibilities considerably beyond the solutions in our case studies, even when rewards and punishments are integrated. It is well established that markets can be designed so that those involved have incentives to act in a coordinated manner suitable for dealing with second-order problems.

For a simple example, consider when we want to influence how much water to consume and from what sources (given different environmental externalities associated with water consumption from different sources). These concerns essentially underlie the *Tampa Bay, East Central Florida*, and *Apalachicola* cases.[7] We can easily imagine adjusting prices so that the consumer pays the true cost of the water utilized, providing correct incentives not only for consumption but for investments to reduce water use, mitigate environmental damage, and develop new sources with less environmental consequences. Also, we could generate funds to develop alternative sources of water supply and means of environmental remediation and prevention.

Even more complicated problems can be dealt with by such means. For example, the World Resources Institute has developed NutrientNet, a market system designed to give those who would otherwise put harmful nutrients into watersheds (a typical second-order problem) the right incentives.[8] If a decision process can be designed to produce an agreed-on aggregate level of nutrients (perhaps reduced over time), we can create a market in which trades can take place, investments can be made, innovation encouraged, and reductions achieved at low cost. Applied to a case such as *Tampa Bay* where nutrients are an issue, a decision process could be designed to agree on a goal (e.g., reducing nutrients so that sea grass—a key indicator of the water's health—grows at 2 percent a year) and a system such as NutrientNet could be implemented.

Yet, for intuitive reasons—voters dislike higher prices even more than inefficient policies or despoiled environments, and will punish politicians who raise prices—market mechanisms are scantly mentioned in our cases. They are dismissed out of hand when considered at all.

For example, consider using prices to resolve water supply problems in *East Central Florida*:

> Ideally, excessive use of a good could simply be corrected with higher prices. This would make the providers keep their levels of benefits while protecting the aquifer at the same time. But elected officials are motivated by the desire of getting reelected. Providing large quantities of the resource at a lower price helps them in reaching this goal. (Berardo 2003)

Similarly, for the *Apalachicola* conflict and water rights:

> The negotiators have steered away from a market valuation of water leaving only a "rational and scientific" assessment of water needs (Water Governance Working Group 2003, *180*).

Despite the intuitive reason for dismissal, we might ask, given experiences elsewhere, if market choices might be adopted.[9] National markets may have been quickly dismissed before their adoption, and yet politicians have lived to see another day. Certainly there are Florida interests, such as the Florida Council of One Hundred (2003)—a group of key state business leaders—who will consider markets. And, certainly, environmentalists in various contexts have been willing to endorse markets (see Chapter 12).

While completely answering such a question requires a more thorough analysis of the political economy of instrument choice (Keohane et al. 1999) than appropriate here, we could consider whether environmentalists and firms, both of whom have much to gain, could be mobilized—perhaps with prodding from national stakeholders as they have done in other instances (e.g., air quality in southern California)—in support of a grand bargain to implement market mechanisms (Hahn 1990). If feasible, theory and evidence suggests that this solution would solve at least a subset of Florida water conflict problems better than a reform of bureaucratic institutions.

We could imagine a successful mobilization entailing a number of features (Stallworth 2000):

- a large "public learning" component showing how polluter-pays principles can be integrated with environmental protection and previous experiences with such mechanisms;
- guarantees that resources generated would be employed for environmental infrastructure, so that those in areas whose water might be diverted (e.g., Northern Florida), or who are less well-off financially, would be assured of tangible benefits; and
- protections for those of lesser monetary means to assure that their minimum needs are met ("lifeline rate structures").

We may not necessarily want to dismiss nonincremental reforms as a means of adaptive governance. We should recognize that eschewing such solutions frequently means throwing out the best means of adaptive governance.

Concluding Observations

We naturally want win-win solutions in which multiple stakeholder interests and the environment are all well served, none at the other's expense. Indeed, some scholars make strong claims that, with the right approach, stakeholders can forge cooperative solutions in which everyone is better off (Porter and van der Lind 1995; but see Palmer et al. 1995).

Our case studies show that the reality of resolving second-order environmental conflicts such as those in Florida is often more convoluted than such discussions admit. A set of political institutions and corresponding choices, coupled with inherently complex public policy issues, virtually assures a messy and complicated process in the search for efficiency and equity.

Rather than a simple roadmap to greater equity and efficiency, our case studies, particularly in light of experiences elsewhere, provide at least two insights. First, as is well recognized, when adaptive governance relies on adding or amending bureaucratic administrative structures, public policy will likely only change incrementally, as the institutional and statutory system is constraining. Quite frankly, adopting more market-oriented solutions where appropriate—if politically feasible—would make a more substantial and benign contribution to realizing the goals of adaptive governance, such as efficient, and hopefully equitable, management of water resources.

Second, when adaptive governance does rely on more conventional administrative structures as engines of change, we need to think what will keep relevant stakeholders committed to the process even when illuminating public and scientific learning occurs. Regardless of artful design processes constructed by those with the best intentions, disappointments will likely result without a credible set of rewards and punishments. Even then, many attempts at improving administrative processes will seem awkward and second-rate. Nonetheless, environmental policies in the United States and other advanced industrial countries are credited with notable achievements, despite their weaknesses, and hopefully administrative innovation in Florida will produce similar outcomes.

Notes

1. Despite conceptual problems, I find that it is easier to discuss whether efficient solutions are adopted rather than equitable ones (except to the extent that fairness involves current stakeholders having their interests accounted for in the decision process). While we might define intergenerational equity as assuring the same quality of water resources enjoyed by the current generation, analogous to Solow's (2000) definition of sustainability, this metric is never explicitly discussed in our case studies. Thus my discussion gives primacy to efficiency.

2. For example, this perspective has been applied to long-term relationships between legislators and organized interests (McCarty and Rothenberg 1996).

3. In another instance of endogenously created interests, the Ocklawaha case (where government actions created recreationalist interests that support the reservoir's existence), deliberation might be more successful if less investment had taken place.

4. As of this writing, the national commitment for restoration appears firm despite budgetary woes and worries about President Bush's environmental commitment, perhaps augmented by the relationship between the president and the governor (Anselmo 2003).

5. The Georgia Pacific case might also be called a success without anything resembling adaptive governance, as the parties use the administrative hearing process and the permit is granted to the firm. The issue is not settled; it is being appealed and may ultimately appear messy and dysfunctional, or the parties may work out an agreement within the adversarial confines of the administrative and judicial systems.

6. Text available at http://gos.sbc.edu/b/browner.html.

7. Of course, there are other complications in these cases, such as disputes over property rights, which require settlement.

8. Information on NutrientNet can be found at www.nutrientnet.org.

9. Admittedly, one difference between simply increasing the price of a good such as water to capture externalities, and the market mechanisms commonly used in the United States, is that consumers feel the former more directly. Although Europeans have willingly used taxes in this manner, American politicians have preferred less transparent quantity mechanisms such as tradable permits.

CHAPTER 22

Conclusions

The Future of Adaptive Governance

Bruce Stiftel and John T. Scholz

THE EIGHT CASE studies and twelve analyses in this volume traverse a wide range of issues pertaining to the five challenges of adaptive governance: representation, deliberative process design, scientific learning, public learning, and problem responsiveness. This final chapter speculates about the future of adaptive governance. It summarizes findings about the five challenges, suggests some characteristics of an institutional framework that might serve the experimental needs of twenty-first-century water policy, and considers the role of adaptive governance in the overall governance of water and other natural resources.

Water governance in Florida at the start of the twenty-first century is highly decentralized, broadly consultative, and innovative. The cases and analyses illustrate just how far Florida water policy has come from the engineering norms of the 1950s and the centralized command-and-control perspective of the 1970s. Confidence in engineered solutions has eroded dramatically, with many of the recent controversies involving the dismantling (*Ocklawaha* and *Everglades*) or reassessment of the use (*Apalachicola)* of structures from earlier decades. The ability of central authorities to impose water policy on passive user communities is a thing of the past: in most of our case studies—*Everglades*, *Ocklawaha*, *Suwannee*, *Fenholloway*, *Aquifer Storage*—user communities would not sit still for proposals that came from USACE, EPA, FDEP, Congress, or the state legislature. The range of agencies responsible for water policy in the state is extensive, with significant authority exercised at all levels of government and in special regional districts.

Procedures for consultation with user groups have developed in several directions. Conflict eventually led to consultation and creative policy innovations in *Everglades* and *Tampa Bay*. *East Central Florida* and *Suwannee* remain promising, although conflict outside the Initiative threatens the *Suwannee* case. *Aquifer Storage* still lacks a clear forum to integrate scientific and public learning in order to develop coherent long-term policies.

The remaining cases document less successful attempts to resolve the problems arising from a fragmented institutional legacy that struggles to keep pace with the pressures of urban and environmental constraints, particularly those related to growth in demand, unexpected responses from natural systems, and scientific progress. Efforts in *Fenholloway* never secured general public representation and collapsed without bringing parties to agreement. No effective decision process was crafted for *Ocklawaha*. *Apalachicola* struggled to integrate science and may wind up back in court.

Representation

Who should be involved in what decisions, with what resources, and with what authority? In an ideal democracy, all citizens have equal access to public policy decisions, but fully participatory processes are in truth costly and cumbersome.

The challenge of representation is least problematic for well-organized competing interests, as is evident in the *Everglades* case. A special Governor's Commission for a Sustainable South Florida, followed by the South Florida Ecosystem Restoration Task Force, provided venues in which the competing interests succeeded in developing several hundred integrated projects, valued at over eight billion dollars, to extend and manage the water infrastructure for the complex urban, agricultural, and conservation interests in South Florida. In the *Apalachicola* Basin, where representation reflected state lines rather than user groups, the challenge was much greater. State negotiating teams struggled unsuccessfully to reconcile the need for public discussions within states to develop an acceptable, coherent state position with the need to maintain secrecy to negotiate with the other states from a position of strength.

Many of the water conflicts involve new interests seeking representation in existing decision processes that have previously excluded them. *Fenholloway*, after painstaking initial successes, eventually failed to involve environmental groups in a process that integrated the traditional permitting system with ecosystem planning efforts. The immediate conflict that evolved into the *Apalachicola* negotiations arose when downstream users in Florida and Alabama sought a voice in the USACE's flow management decisions for the dams it had built for power and water supply upstream. Ruhl argues that agencies sometimes thwart the representation of affected interests; they may be oriented toward a fixed group of clients, their outlook may be restricted, or the affected interests may not be organized to participate in existing agency decision processes. Filtering input through multiple layers of nested political institutions, as in *Apalachicola*, can effectively silence constituencies. Adaptive government efforts frequently exhibit the tension between participation of diverse, potentially divisive groups and a cohesive decision process capable of developing and implementing plans.

Forester picks up Ruhl's point, arguing that when those affected by decisions are disorganized and even unaware of future implications of decisions, it is often impractical for agencies to involve them. *East Central Florida* describes the problems facing the water management district in its attempts to involve all relevant water utilities, and particularly in maintaining the interest of elected officials who would be asked to fund joint projects. The *Ocklawaha* and *Suwannee* initiatives both failed in their attempts to involve groups who in the end opposed the decisions, through the legislature in the first case and the courts in the second. In *Aquifer Storage*, many affected groups were absent from technical policy debates; most were drawn in only as last-minute opponents when supporters sought legislative approval for a policy that had been shielded from public debate.

Even successful mobilization may not guarantee participation. Insufficient technical capacity may stand in the way, as argued by Ruhl and by Susskind. Free-rider dynamics may come into play when individuals, firms, or groups rely on others to speak for their interests, as suggested by Lubell. *Aquifer Storage* evidences the considerable expense of technical representation that exacerbates representational bias in technical forums. Susskind argues that those who need technical assistance in order to participate should have it provided for them, and that effective preparation of stakeholders is a responsibility of the convenors, not just the stakeholders themselves. Lubell argues that involving governmental authorities who are trusted by key groups can make the difference between free riding and active participation; in his view, free riding can be reduced by establishing high expectations about the participation of others in the stakeholder group. If all else fails, Forester suggests surveys and other devices to incorporate the concerns of absent stakeholder groups.

Finally, the long time horizon of many adaptive governance issues imposes a critical representational challenge during the implementation processes. Jones notes that "successful" adaptive governance processes will inevitably discover problems with some assumptions behind the initial agreements. Maintaining consensus despite changing knowledge poses a great difficulty, particularly after initial leaders and representatives have moved on to other issues or other stages of their careers. Jones argues that leadership is the critical element in overcoming this problem, but an enduring system of representation capable of sustaining engagement of groups as projects unfold may prove more reliable if such a process can be developed.

Which standards of representation are most appropriate for adaptive governance processes depends on the circumstances in which representational challenges arise. The more that a process meets standards of representation acceptable to the greater political community, the less likely the outcomes will be opposed administratively, legislatively, or judicially. As Rothenberg points out, the fragmented American federalist system provides ample opportunities for opposition to halt innovative policies, as exemplified by *Ocklawaha* and *Suwannee*. While these present obstacles to efficient policies, they provide a powerful incentive for adaptive governance to ensure representation of those interests capable of appealing to political and judicial authorities. Quirk, on the other hand, criticizes Florida's water management structures because they fail to ensure fair representation. He believes that the ad hoc collaborative processes normally associated with adaptive governance are by their nature undertaken by authorities that do not have access to the complete scope of affected interests. In his view, nothing short of new state institutions that provide access to all constituencies will meet the representational needs of the state.

Process Design

Adaptive governance must elicit a reasonable understanding of what represented groups prefer, translate those preferences into policy, gain necessary approvals, and assemble sufficient resources and skills to implement the preferred policies. Each of these requirements can be difficult to meet.

Even if all stakeholder groups are involved, representatives may not fully understand and articulate their interests. Ruhl argues that participants may not understand the nature of the problems, the viable options, and even the needs and values of their own constituencies. So groups often lump all demands together without prioritizing the underlying objectives, and the resultant "group think" naturally leads to exaggerated demands. Forester finds that parties' claims may be logically or dramatically contradictory, often resting on assumptions that would not hold up if critically examined. *Fenholloway*, *Ocklawaha*, and *Apalachicola* provide examples of decision processes that failed when prematurely-articulated interests were never clarified through subsequent analysis.

Translating interests into acceptable policy options is the core of the collaborative process, but is often difficult because of the adversarial nature of the regulatory system. Sabatier describes how coalitions view opponents as more nefarious and more powerful than they really are, leading to greater solidarity within the coalition but greater conflict with outside groups. Forester also analyzes the institutional propensity to resist learning from other groups; believing that misrepresentation can be to their strategic advantage, stakeholders may exaggerate and posture rather than a search for common interests. In the chronic water wars in *Tampa Bay*, ad hoc consensual processes were pointless as long as opponents persisted in the self-defeating suspicion that others were using these processes for strategic advantage.

Faced with this frustrating dialogue, agencies and leaders press experts for fixes before they can understand the issues and interests, and the resulting solutions may not be sustainable. Jones calls this a "flight to authority," exemplified by *Apalachicola* and *Aquifer Storage*. Leaders in *Everglades* initially pressed for a hasty solution, but later committed to reach the understanding necessary for an enduring solution. Technical assistance to interest groups may encourage them to persevere and understand complex underlying issues, but such assistance was rare among the cases, and seldom came from disinterested sources. The water management districts often provided technical analysis, process consultants and their subcontractors were less often involved, and university scientists were only sometimes drawn in. In a few of the cases (*Everglades*, *Tampa Bay)* resources for technical support were appropriated independently, but in most (including *East Central Florida* and *Aquifer Storage*) technical support was volunteered by one or more of the involved parties, raising doubts about objectivity.

As suggested by Susskind, neutral facilitators can minimize several of the problems discussed here. Neutrals can assess conflicts prior to formal discussions, helping to ensure the involvement of all affected groups with sufficient technical support. As Forester argues, neutrals can push for ground rules that minimize misrepresentation, and can show groups how a narrow focus on demands work against them. Neutral process facilitators were involved in many of the cases, but often under less than ideal circumstances. With the exception of *East Central Florida*, formal conflict assessments seldom took place. Discussions of ground rules were sometimes perfunctory. In a few instances, facilitative processes paralleled ongoing legislative dialogues that overshadowed them (*Aquifer Storage, Apalachicola*). In contrast, when professional standards of neutral process were largely adhered to (*Suwannee, Tampa Bay*), solutions were reached that seem more likely to endure.

It is not surprising that many of the cases suffered these problems of process design. These ad hoc collaborative processes are relatively new, often misunderstood, and sometimes without clear statutory authority. For comparison, consider the expansion of rulemaking and adjudicatory powers of administrative agencies that over the past century have become a critical part of the national system of governance; procedures developed for a particular problem eventually became codified in the federal Administrative Procedures Act and legitimized through a body of court decisions. If adaptive governance is to become a commensurate force in policymaking in water and other environmental areas, the role of adaptive processes will need to be clarified, the requirements codified, and the extent of authority determined.

At minimum, new decision processes must be compatible with administrative, legislative, and judicial authorities that otherwise can nullify agreements. Several of our cases began with court cases or judicial appeals (*Everglades, Apalachicola*) that were stayed pending the attempt to develop agreements through adaptive governance. All of them involved administrative agencies, generally from multiple levels with authority over diverse administrative domains, which generally signed Memoranda of Understanding to ratify agreements reached within the processes. And often, as in *Tampa Bay* and *Ocklawaha,* ratification by the state and or federal legislatures was necessary.

The problem of deviating from the agencies' existing routines without agreement from all parties is illustrated by the continuing saga of the *Suwannee* Initiative, where a court challenge by environmentalists who did not participate has required FDEP to proceed with formal permitting despite the voluntary best management practices implemented by affected users. This demonstrates the advantage of obtaining legislative ratification of locally derived agreements, as in *Tampa Bay* and *Everglades*—elected politicians generally have little problem with accepting credit for passing requested legislation once all involved groups are in agreement. On the other hand, in *Aquifer Storage* proponents failed to get legislative approval to expedite developments, a sign of the difficulty in gaining legislative backing for ad hoc projects, particularly when opposition groups are not involved in earlier proceedings. Generic authorization of these processes, along the lines of the Administrative Procedures Act, would help ensure the implementation and enforcement of agreements.

Susskind, in particular, advocates the establishment of consistent national guidelines to govern decision processes, although he appears to envision something more akin to professional guidelines than a legal code. This would leave more flexibility to design processes to fit the conflict being addressed, and consensus on policies derived in conformance with guidelines are unlikely to be rejected by statutory and constitutional authorities.

Rothenberg is more concerned with creating appropriate incentives. He suggests that successful agreements depend on credible rewards and punishments to nudge actors toward compliance. To provide such credibility, Quirk emphasizes the need for a designated agency to develop consistent guidelines, oversee the process, and validate decisions reached through collaborative processes. Providing a clear institutional framework is particularly important for long-term adaptive processes, as noted previously, to maintain support even as scientific and public learning compel alterations in initial agreements. A critical task of this agency would be to develop a secure system of accountability that will reassure participants that agreements will be implemented yet allow the flexibility to adjust to new scientific and public learning.

Scientific Learning

The case studies show that integrating science into decision processes involves disagreements among scientists from different disciplines, insufficient or unbalanced access to science by stakeholders, and the limited capacity of scientists to provide what policymakers expect. In *Everglades*, biologists studying a particular species or ecosystem were frequently concerned with different questions than were other scientists and the officials responsible for managing the habitat. In *Fenholloway*, the standard of scientific proof was itself in dispute. In *Suwannee* and *East Central Florida*, science was uncontested, perhaps due to the disjuncture between science and the concerns of the key parties. In *Aquifer Storage*, the Florida legislative debate reflected a preference for science that would justify policy rather than policy based on science. In the *Ocklawaha* we see the futility of long periods of study that neglected to engage one of the critical parties, since that party subsequently undermined the scientifically supported agreement through legislative action.

Science is inherently about disagreements and the standards to be used in settling them. Traditionally, scientific review by "neutral" peers provides the norm in judging evidence. Yet water and other environmental conflicts about science may call for different standards for two reasons. First, dialogues frequently involve different disciplines that must not only develop a mutually respected set of peers but resolve terminological and procedural differences for the dialogue to begin. The frustrating dialogues between hydrologists and biologists in *Apalachicola* provide a typical example. Second, Sabatier argues that scientific disagreements inevitably become embedded in disputes among competing users and authorities, each with a different notion of which scientific discipline and which experts are most relevant. Indeed, Ozawa warns that available scientific knowledge can sometimes aggravate conflict and prematurely narrow the potential range of policy solutions.

Linking scientific learning with policy processes is complicated by the mismatch between what policymakers and users expect from science and what science is prepared to provide. Ozawa explains that policymakers often want answers that science cannot provide. These "trans-scientific" questions include issues for which appropriate data have not yet been assembled, questions that are outside the domain of science, and questions on which there is still disagreement within the scientific community. According to Susskind, this uncertainty provokes frustration—scientists can name possibilities, but can't assign definite probabilities to outcomes of proposed courses of action. Targeted research may clear up disagreements or lack of data, but cannot resolve policy disputes beyond the current abilities of science, when the necessary basic research will exceed the resource capabilities and time frame of most decision processes. Sabatier notes that disagreements among the social sciences about responses of the human system to policy alternatives are generally the most difficult to resolve.

Given this mismatch, Ozawa and Sabatier both warn against attempts to insulate the scientific learning process from partisan interests, since doing so is unlikely to lead to consensus. An advisory board of scientists isolated from the policy disputes in *Tampa Bay* could agree on a final report, but the report had no apparent impact on the water wars. Ozawa emphasizes the importance of involving partisans early in the process of determining what knowledge is relevant, which experts should be involved, and which studies are needed to resolve scientific issues behind policy disputes. Without early involvement, local knowledge from user groups will not inform the scientific inquiry, and important user issues may be overlooked by scientists and administrators concerned with their own issues and policy preferences. Unless all relevant policy coalitions are represented by trusted experts, Sabatier tells us, scientific committees are unlikely to develop consensual scientific foundations for policy decisions. Ruhl, Susskind, and Ozawa all call for methods that make technical support available to groups that cannot provide it on their own. Forester adds that neutrality in such support is key, and calls for universities to be brought in.

The combination of scientific uncertainty and partisan interests lead Sabatier to emphasize the importance of designing an appropriate forum for long-term dialogues, one that includes partisan experts but is dominated by prestigious "neutral" scientists capable of imposing scientific criteria on findings. In complex, long-term projects like the Everglades that involve considerable scientific uncertainty, institutionalized review linked directly with ongoing project reviews can maintain a consensus as the scientific basis of initial agreements evolves. Susskind notes that this institutionalized process of designing policy as an experiment and adapting policy to enhanced knowledge is the primary goal of adaptive management, a goal that policymakers seldom fully realize.

Public Learning

Scientific learning cannot resolve conflicts without being linked to public learning about the causes, consequences, and potential solutions of the new water conflicts. Yet this critical goal of adaptive governance remains the least understood and most ambiguous challenge. For stable implementation, widespread public support is often vital to adaptive government agreements; in the short run, it may be needed to force compromises and gain compliance with policy decisions, and in the long run it may discourage legislative and judicial challenges. Yet such public involvement, even involvement of members of key water stakeholder communities, is rare and, when it does occur, unlikely to change attitudes or behaviors. Thus our cases and analyses present only a limited number of issues involving the public learning process.

Almost all the cases were the subject of considerable media coverage, especially *Everglades* and *Apalachicola*. Even localized conflicts like *Fenholloway* entered the national limelight for a time. But media coverage rarely translated into widespread notice by disinterested members of the public. Stakeholder communities were represented in the deliberations of *Suwannee* and *Everglades*, but in many of the cases, including *Ocklawaha*, *Aquifer Storage,* and even the nationally televised *Fenholloway* debate, stakeholder involvement was limited to small groups of highly motivated individuals. In *Apalachicola*, there were extensive efforts to involve specialists, but lay stakeholders were kept at a distance—often, it seemed, as a strategic decision to keep control in the hands of leaders who wanted bargaining flexibility, including the choice of which interests to trade away. Both Quirk and Rothenberg fear that such dismissal of constituencies may not be rare.

Process design can influence the public learning process, particularly among participating user groups. Lubell shows that the involvement of local government and agricultural agencies in *Suwannee* was critical to enlisting farmers in voluntary programs to reduce runoff. Trusted officials persuaded prominent members of the farming community to participate, which subsequently influenced others. Roberts shares Lubell's view that success in *Suwannee* hinged on the involvement of many farmers, and emphasizes that individualized incentives made that happen. Both would agree with Sabatier's warning that incentives alone are unlikely to be sufficient; authorities who do not pay sufficient attention to stakeholder belief systems in program design will lose participation and compliance. In short, the involvement of trusted officials in *Suwannee* provided a link to the user community whose behavior was to be changed, ensured that resources for incentives were available, and increased the congruence between the program and farmers' belief systems.

Roberts emphasizes that learning takes place among stakeholders directly and indirectly through participation, but Sabatier cautions that learning across advocacy coalitions is difficult. According to Sabatier's Advocacy Coalition Framework, public learning takes place primarily within longstanding coalitions as allies refine their beliefs over long series of political battles. He suggests that core values of the members of these coalitions change more slowly than policy-specific attitudes and beliefs, and that considerable time and unusual circumstances are required to change even those. "Hurting stalemates" may be necessary before policy opponents will even participate in processes that might change at least their policy-specific attitudes. The *Tampa Bay* water wars took decades to reach a sustainable compromise, and did so only after key initial participants with entrenched positions dropped out. The Tampa Bay Authority did not end conflicts, but rather established a consensus on the critical issues and the appropriate means to resolve differences; this allowed long-delayed projects to move forward.

Rothenberg emphasizes the need for sufficient incentives to induce changes in policy positions among participants, particularly for groups served by the status quo. *East Central Florida* provides an interesting but unfinished test of whether the *Tampa Bay* example can stimulate learning in a similar policy arena that has not directly experienced the hurtful stalemate. After all, the water management district responsible for initiating discussions is threatening local water suppliers with restrictions on groundwater pumping permits; these eventually brought a resolution in *Tampa Bay,* but only after numerous legislative and judicial challenges. Rothenberg and Quirk both note that consensual processes depend ultimately on whether state authority will support agreements when challenged, and that clear policies and institutions establishing the legitimacy of such agreements are necessary to motivate conflicting parties to participate meaningfully in processes with the potential to alter beliefs and stimulate public learning. To ensure that actual practices in adaptive governance match the expectations on which such legitimacy is based, Quirk emphasizes the need for an oversight agency to review procedures and to act as arbiter. Only within a clear context of legitimate authority are the learning processes discussed in the sections on representation and process design likely to emerge.

Problem Responsiveness

These cases can make only a weak argument for enhanced efficiency and greater fairness. *Ocklawaha* and *Aquifer Storage* have yet to reach outcomes. In others, including *Everglades* and *Tampa Bay*, measurements of success are elusive. But documenting success is difficult, in part because the appropriate standards of efficiency and equity are not clear.

East Central Florida is a case in point. Berardo describes the reaction to the water management district when it rewarded utilities that conserved water by increasing their pumping limits. Other utilities viewed these higher limits as arbitrary and unfair, since there was no consensus on the relative responsibilities for conservation and little transparency in the permit renewal process. Our analysts are critical of the outcomes in cases where they can be assessed, sometimes disagreeing as to why. Hamann and Susskind find a status quo bias that makes change hard. Rothenberg decries the snail pace of progress. Pollman, on the other hand, decries the painful and costly periods of instability that prevent investment in infrastructure. Rothenberg echoes concern with climbing costs, and Hamann laments that when change occurs, it is often at public expense. Susskind finds little movement toward sustainability or truly adaptive management.

Some of the criticisms are aimed at the adaptive processes themselves. Rothenberg finds that stakeholder desires are increasingly at odds, and that the choices made are often inefficient. He is particularly troubled that market incentives are hardly ever considered. Quirk sees an overall inability to cope, evidenced by the state's fragmented system.

We can argue that the voluntary nature of collaborative processes can lead to resolutions that should be Pareto-superior to those imposed by administrative fiat. However, the long-range contribution of adaptive governance to the American federalist system remains an important area for study.

Improving Adaptive Governance: Stronger Collaboration, Pragmatic Science, Wiser Competition

Three key themes for improving adaptive governance have emerged: stronger collaboration in consensual processes, more realistic use of scientific information, and greater incorporation of market incentives.

Stronger Collaboration

Forester, Jones, Lubell, and Susskind call for improvements in the way we conduct collaborative policymaking. We need to select the best personnel for consensual processes: stakeholder representatives who embrace leadership, and properly qualified neutrals. Surrogates should be used only when constituencies cannot or will not name representatives. Neutrals, not domineering agencies, should steer the process. Ground rules should be worked out in advance. Assessments of the views of the parties must be clear and specific.

Developing a formal, well-considered, uniform standard for making interagency consensual water policy could extend the use of adaptive processes. Formal guidelines could give participants clearer expectations, reduce startup costs, better match the decision process to the type of conflict, restrict the process to viable situations, and provide accountability in subsequent judicial, administrative, and political reviews.

Since standardization would reduce the flexibility critical to adaptation, guidelines must be tailored to various categories of conflicts, and should not preclude experimental procedures in special circumstances. Collaborative leadership training should encourage creative management within the formalized guidelines. For adaptive governance to play a large role in the overall governing process, a specialized state authority may be necessary to create formal rules as well as to develop oversight mechanisms and appeals processes that would legitimize decisions and moderate opportunism. The specialized authority can support public and scientific learning, providing extra safeguards from intervention by legislative, executive, and judicial systems.

Pragmatic Science

Ozawa, Roberts, Sabatier, and Susskind all call for better ways to encourage scientific and public learning during water policy discussions. Unrealistic expectations of neutrality should be amended to recognize the partisan nature of scientific debate itself, and the divergence in goals between science and policymaking has to be acknowledged. Partisan interests need to be represented in designing research to resolve policy disputes, but the scientific forum needs neutral leadership to ensure that scientific peer review standards guide the evaluation process. Independent scientific consultants hired as staff can provide a counterbalance to those invited by advocates, particularly to overcome imbalances in the expert resources of different groups. Substantial time is required to form a consensus on critical issues and to extend this consensus beyond the experts as part of the public learning process. Preliminary decision processes need to be formalized in order to monitor outcomes, evaluate changes in scientific understanding, develop consensus around these changes, and legitimize adjustments in policy.

Wiser Competition

Polmann, Quirk, Rothenberg, and Ruhl are skeptical that improvements in collaboration or in the use of science can meaningfully advance water policy. They observe that the fragmented governance system strongly favors the status quo over proposals for change, that the cost structure of water in Florida greatly undervalues the worth of the resources, and that fundamental changes are necessary to reconcile demand with supply. Moreover, they see stability and predictability as important values that may be undermined by inappropriate adaptive governance procedures.

Among the suggestions is a call to enhance rewards and punishments to motivate actors to move toward agreements and comply once agreements are made. Given the costs and limitations in developing consensual policies, Rothenberg in particular advocates market mechanisms to adapt to changing circumstances. To the extent that incentives can align demand with supply, money is more likely to be invested in developing alternative supply and conservation rather than in the permit battles illustrated in the *Tampa Bay* case. Incentive-based policies that allow adjustments without policy changes can extend the capabilities of adaptive governance, and can ensure that investments in adaptive governance are aimed at issues for which incentive-based systems are unsuitable.

The challenges of developing market-based approaches are considerable, particularly in guaranteeing equity across user types, maintaining the adaptive value of prices, and ensuring the continued integrity of the ecosystem through monitoring systems directly linked to price mechanisms. Ultimately, adaptive governance institutions may be most successful as governing bodies overseeing market-like systems, monitoring the expected behavior of both the market and natural systems, adjusting incentives when necessary, and adapting the system to unexpected results. Despite the dual challenges posed by the widespread belief that water is a "right," not a commodity, and the political unsavoriness of raising prices for something that all voters must use, experiments with market mechanisms among competing environmental users deserve study.

Contributions to Governance

Our final task is to consider the overall role of adaptive decision processes in the governance of water and other natural resources. We discuss first the contributions that could be expected if adaptive governance processes achieve their potential, and then the limitations of expanded use of these processes.

The major contribution arises from the consensual nature of the process, in marked contrast to the adversarial legal system and the opposing advocacy coalitions in the political and administrative arenas. Given the complexity, scientific uncertainty, and unexplored policy alternatives involved in the "second-order" conflicts that concern us, the ability to bring opposing sides together allows for the joint exploration of a considerably larger search space to find mutually advantageous policies. Extensive search is particularly appealing in situations where knowledge is minimal and affected users may scarcely understand their long-term interests. To the extent that scientific exploration and policy negotiations are successfully integrated, adaptive governance enhances the likelihood of achieving a consensual basis for an adaptive policy process.

Several related contributions follow from applying consensual methods to the limited geographic scale and narrow scope of most adaptive governance procedures. First, public and scientific learning processes can most readily be linked within a geographically defined natural system. The limited set of users can play a more direct role in considering the policy alternatives and the scientific agenda necessary for evaluating those alternatives. Stable local representation can learn about relevant scientific findings, and credibly share new information with their constituents. Local representation within a consensual process can thus provide the political basis for responsive management.

Second, specialized processes can design policies based on the specific characteristics of the natural and human systems. Given the variations in both systems, a uniform national or even state policy that is globally "optimal" will usually be locally suboptimal. Locally-designed policy variances can take advantage of local characteristics. Local resolution of conflicts can thus enhance the efficiency of national policies without making national standards more complex.

Third, institutional, integrated public and scientific learning combined with local policy variances can adapt in ways impossible at the national level, where decisions once made tend to be difficult to change. Adaptive governance processes act on a scale most likely to provide an experimental laboratory for the exploration of new policies—a traditional role of decentralized processes in federalist systems, but one that existing local government jurisdictions have difficulty fulfilling for complex and contentious water conflicts. Once policies have proven sustainable in one area, they can be extended to other locales. Furthermore, local consensus may engender coalitions to extend the policy on the state and national level.

Fourth, specialized processes involving local governments could integrate land-use policies traditionally controlled by local governments with environmental policies traditionally developed at the federal and state levels. Adaptive governance institutions can thereby extend the range of potential policy solutions beyond what have generally been available.

Adaptive processes also increase the social capital that can contribute to the resolution of issues. The cooperative working environment diffuses knowledge about individuals, organizations, logic, and resources that can be used to prevent or manage future difficulties. Successful collaboration can transform the way actors look at institutions, and the motives they attribute to both the members of those institutions and their adversaries. A better understanding of collaborative processes and the issues at stake allows participants to address critical challenges to adaptive governance before they disrupt the process.

Finally, adaptive processes alter perceptions of justice. When people deliberate in a fair and open process, they naturally tend to accept the result. Ideology and brute self-interest cannot help but be muted in the face of opponents who are seen to be working hard to find a way to solve everybody's problems.

Limitations

While appropriate adaptive processes and institutions can contribute a great deal to the resolution of second-order conflicts, several factors ultimately limit the role they can play within the overall governing system.

Participants may fear that adaptive processes will augment the influence of powerful constituencies on government agencies. Adaptive processes create venues where mobilized stakeholders can press agency officials for favorable decisions, while less organized or geographically remote groups are less likely to have comparable access. Well-designed processes, guidelines, and oversight may guard against such inequities, but adaptive governance processes will not overcome participants' fears without considerable effort.

Successful negotiations and ad hoc institution building consume the time and money of those involved, and potential gains from agreement depend on the cooperation of opposing groups. As the cases and analyses have shown, all parties must face a "hurtful stalemate" before they are likely to initiate a process capable of compromise. Successful negotiations take time to establish trust and instill science-based public learning. Institution building and maintenance requires active leadership, and adequate representation requires a stable group able to devote sufficient time and energy.

When the required factors come together, adaptive governance can contribute a great deal both to the resolution of the regional conflict and to the development of new policies useful in other locations. Whether successful models can work in less fortunate situations remains unclear, however, since lessons learned in showcase situations may not be relevant when fewer resources are available. Government agencies can reduce this limitation by investing resources in the establishment of adaptive governance institutions, as many state and federal agencies have already done. However, current efforts channel resources into successful showcase efforts and away from problematic areas less able to resolve their own problems. Extending adaptive governance requires a better understanding of startup problems in areas with less social capital.

In particular, the linking of scientific and public learning is constrained by limited resources. Local resources available to specialized adaptive processes are insufficient to provide basic scientific research, particularly in developing new theories and new research tools. Showcase megaprojects like *Everglades* are likely to uncover new knowledge relevant to smaller adaptive efforts, but in general local decision processes can only apply known research. The extent to which megaprojects will provide scientific knowledge useful to average, resource-constrained projects remains to be seen.

Of course, federal and state agencies could devote more of their research and implementation budgets to overcome local resource constraints, but they understandably tend to maintain their autonomy from other institutions and limit the discretion granted to adaptive governance. As Pollman notes, adaptation can impose delays, uncertainties, and other costs on all activities, from planning through the development and maintenance of infrastructure, to the enforcement of rules of use. The tradeoff between managerial efficiency and adaptive ability is particularly problematic; processes designed to encourage participation are exploited by users seeking greater advantages than they could achieve in existing agency decision processes, thereby reopening previously settled disputes in user communities.

At best these impediments take up agency time and resources, while at worst they may undermine the agency's ability to function, particularly if groups that traditionally influence the agency differ substantially from those recruited into the newly developed adaptive governance institutions. In short, existing authorities are likely to resist the delegation of broad powers to local, ad hoc processes whose results are unpredictable. The problem of delegation currently constrains the scope of agreements that can be reached by adaptive governance to relatively minor variations within existing policies—greater variation requires federal or state legislative approval. Accountability mechanisms developed through uniform guidelines, statutory requirements, or an oversight agency may expand this scope for experimental development, but local variations granted to adaptive governance institutions can only go so far in resolving more basic problems in the federal and state legal framework.

Summary

Adaptive governance institutions can play a significant but limited role in resolving the new conflicts that confront the patchwork of federal, state, and local agencies. They can enhance overall policy by coordinating the policies of fragmented authorities. They can change relations among adversaries and promote joint problem solving, initially in the specific locations where institutions have emerged, but potentially at higher levels of government as well. They induce participating user communities to support new policies. They provide cauldrons of innovation for developing new policy ideas and expanding knowledge that can change state and national policies. They tackle conflicts that range from a small river or bay of concern only to local residents to large river systems encompassing multiple states. In doing so, they expand our policy roadmap for adjustments in state and federal institutions and policies.

Adaptive governance processes provide systematic adaptive capabilities that lie somewhere between those of markets and those of democratic institutions. Market institutions encourage entrepreneurs to develop the knowledge and design for new products that meet consumer's needs. Democratic institutions, on the other hand, encourage policy entrepreneurs to use negotiation and persuasion to change voters' perceptions in order to create a new policy consensus. Adaptive governance institutions attempt to link the development of knowledge to the exploration of new perceptions of interests.

Nobody expects these processes to fully resolve conflicts. Disagreements often last throughout the proceedings and live on after agreements are forged. But where they work well, these processes create spaces where adversaries can explore together and develop agreements that leave them better off. Science advances; solutions emerge; but conflict lives on. Only now, it does so with new social and political rules and structures that encourage more efficient and perhaps more equitable next steps.

References

Ackerman, Frank, and Lisa Heinzerling. 2004. *Priceless: On Knowing the Price of Everything and the Value of Nothing*. New York: The New Press.

Adler, Peter S. 2000. *Natural Resources Conflict Resolution: Water, Science, And The Search For Common Ground*. 1st Australian Natural Resources Law & Policy Conference Canberra, Australia (March 27). http://www.mediate.com/articles/adler.cfm.

Adler, Peter S., Robert C. Barret, Martha C. Bean, Juliana E. Birkhoff, Connie P. Ozawa, and Emily B. Rudin. 2001. *Managing Scientific and Technical Information in Environmental Cases*. Tucson, AZ: U.S. Institute for Environmental Conflict Resolution and the Western Justice Center Foundation.

Agreement in Principle on Nutrient Management in the Suwannee River Basin (Agreement). 1999. http://mysuwanneeriver.com/resources/agreement.pdf (accessed February 16, 2004).

Andreu, Ray, and David C. Weeden. 1996. The Fenholloway Project: Reviving a River, Sustaining an Ecosystem, Building Partnerships. *Proceedings of the 1996 NCASI Southern Regional Meeting*, 109–115. Research Triangle Park, NC: National Council for Air and Stream Improvement, Inc.

Andrews, Clinton. 2002. *Humble Analysis: The Practice of Joint Fact-Finding*. Westport, CT: Praeger.

Angelo, Mary Jane. 2001. Integrating Water Management and Land Use Planning: Uncovering the Missing Link in the Protection of Florida's Water Resources? *University of Florida Journal of Law and Public Policy* 12: 223–249.

Anselmo, Joseph C. 2003. Fiscal 2004 Appropriations, Energy and Water Development: Plenty of Debate Expected on Projects. *Congressional Quarterly Weekly Report* June 14: 1441.

Appelbaum, Stuart J. 1999. Telephone conversation, July 6.

Argyris, Chris, and D. Schon. 1978. *Organizational Learning: A Theory of Action Perspective*. Reading, MA: Addison-Wesley.

Arnstein, Sherry. 1969. A Ladder of Citizen Participation. *Journal of the American Institute of Planners* 35: 216–224.

Arthur, Jonathan D., James B. Cowart, and Adel A. Dabous. 2001. *Florida Aquifer Storage and Recovery Geochemical Study: Year Three Progress Report*. Open File Report 83. Tallahassee, FL: Florida Geological Survey.

Axelrod, Robert. 1984. *The Evolution of Cooperation*. New York: Basic Books.

Bahr, Jean. 2003. GSA Birdsall-Dreyfuss Lecture held at Florida State University by Jean Bahr, CROGEE-member. February 24.

Ballard, Bob. 2002. Official Florida DEP (Deputy Secretary) Correspondence with Marsha Kearney, Forest Supervisor, USDA/FS. National Forests in Florida Regarding Special Use Permit for Occupation of National Forest Lands. July 19.

Banister, Beverly H. 2001. Letter from Beverly H. Banister, EPA Region IV Director of Water Management Division, to Mimi Drew, Director of Water Resources Management, FDEP, March 7.

Bardach, Eugene. 1998. *Getting Agencies to Work Together: The Practice and Theory of Managerial Craftsmanship.* Washington, DC: Brookings Institution Press.

Barke, Richard, and Hank Jenkins-Smith. 1993. Politics and Scientific Expertise: Scientists, Risk Perception, and Nuclear Waste Policy. *Risk Analysis* 13(4): 425–439.

Barnes, Nancy. 2003. Personal communication between Nancy Barnes, poultry farmer, and Aysin Dedekorkut, January 29.

Baron, David P. 2001. Private Politics. *Journal of Economics & Management Strategy* 12: 31–66.

Barry, Rick. 1996. Clarke vs. Rainey: Have a Heart. *The Tampa Tribune,* September 26.

Beatley, Timothy. 1994. *Ethical Land Use: Principles of Policy and Planning.* Baltimore: Johns Hopkins Press.

Berardo, Ramiro. 2003. Seeking "Water Peace": the East Central Florida Regional Water Supply Planning Initiative. Tallahassee: Florida State University, DeVoe L. Moore and Family Center for the Study of Economic Policy and Government, April.

Bernstein, Jacob. 2000. Deep Well Infection. *Miami New Times*, Oct 5. http://www.miaminewtimes.com/issues/2000-10-05/feature2.html/printable_page (accessed May 19, 2003).

Berry, Jeffrey M. 1999. *The New Liberalism: The Rising Power of Citizen Groups.* Washington, DC: Brookings Institution Press.

Bianco, William T. 1994. *Trust: Representatives and Constituents.* Ann Arbor: University of Michigan Press.

Bingham, Gail. 1986. *Resolving Environmental Disputes: A Decade of Experience.* Washington, DC: Conservation Foundation.

Bisbal, Gustavo A. 2002. The Best Available Science for the Management of Anadromous Salmonids in the Columbia River Basin. *Canadian Journal of Fisheries and Aquatic Sciences* 59: 1952–1959.

Blake, Nelson Manfred. 1980. *Land into Water—Water into Land: A History of Water Management in Florida.* Tallahassee, Florida: University Presses of Florida.

Blomquist, William, Edella Schlager, and Tanya Heikkila. 2004. *Common Waters, Diverging Streams: Linking Institutions and Water Management in Arizona, California, and Colorado.* Washington, DC: Resources for the Future.

Boffey, Philip. 1975. *The Brain Bank of America.* New York: McGraw-Hill.

Borkowski, Winston K. Sr. 2003. Personal communication between Winston K. Borkowski Sr, assistant general counsel, FDEP, and the author. June 24.

Boswell, Michael R. 2000. Redefining Environmental Planning: Evidence of the Emergence of Sustainable Development and Ecosystem Management in Planning for the South Florida Ecosystem. PhD diss., Florida State University.

Bowman, James S., and Jill Toa. 1993. Fenholloway Furor: River of No Return? *The Jerry Collins Case Studies in Florida Public Administration,* Case Study Number II. Florida Center for Public Management, Publication Number 81094. Tallahassee: Florida State University.

Bronson, Charles. 2003. *Florida Agricultural Water Policy.* Florida Department of Agriculture and Consumer Services. http://www.floridaagwaterpolicy.com (accessed February 26, 2004).

Brookes, Ralf. 2001. Desal Legal Update: SOBAC has 10 Days to File Exceptions; 30 Days to Appeal Desal Decision. *SOBAC Newsletter* 2(5): 3–4.

Buchanan, James M., and Gordon Tullock. 1962. *The Calculus of Consent.* Ann Arbor: University of Michigan Press.

Burt, Ronald S. 1992. *Structural Holes: The Social Structure of Competition.* Cambridge: Harvard University Press.

Bush, J.E. 2001. Letter from Florida Governor J.E. Bush to President George W. Bush, January 22.

Butterworth, Robert. 1996. Opinion on the Restoration of the Ocklawaha River and Removal of Rodman Dam. Florida Attorney General's Office, February 12.

Carpenter, Susan, and W.J.D. Kennedy. 1988. *Managing Public Disputes*. San Francisco: Jossey-Bass.

Carter, Luther J. 1974. *The Florida Experience: Land and Water Policy in a Growth State*. Baltimore, MD: Johns Hopkins University Press.

Caulfield, Henry, Jr. 1978. Policy Goals and Values in Historical Perspective. In *Values and Choices in the Development of the Colorado River Basin*, edited by D.F. Peterson and A.B. Crawford. Tucson: University of Arizona Press.

Chrislip, David D., and Carl E. Larson. 1994. *Collaborative Leadership: How Citizens and Civic Leaders Can Make a Difference*. San Francisco: Jossey-Bass.

Christensen, Norman L., Ann M. Bartuska, James H. Brown, Stephen Carpenter, Carla D'Antonio, Robert Francis, Jerry F. Franklin, James A. MacMahon, Reed F. Noss, David J. Parsons, Charles H. Peterson, Monica G. Turner, and Robert G. Woodmansee. 1996. The Report of the Ecological Society of America Committee on the Scientific Basis for Ecosystem Management. *Ecological Applications* 6(3): 665–691.

Clark, Ray, and Larry Cantor (eds.). 1997. *Environmental Policy and NEPA: Past, Present and Future*. Boca Raton, FL: St. Lucie Press.

Clarke, Gilliam. 1996. Propaganda Doesn't Solve Water Problems. *St. Petersburg Times*, October 25.

Coglianese, Cary. 2000. Is Consensus an Appropriate Basis for Regulatory Policy? In *Environmental Contracts: Comparative Approaches to Regulatory Innovation in the United States and Europe*, edited by Eric W. Orts and Kurt R. Deketelaere. London: Kluwer.

———. 2003. Is Satisfaction Success? In *The Promise and Performance of Environmental Conflict Resolution*, edited by Rosemary O'Leary and Lisa Bingham. Washington, DC: Resources for the Future.

Coglianese, Cary, and Laurie K. Allen. 2003. Building Sector-Based Consensus: A Review of the EPA's Common Sense Initiative. Working Paper RWP03-037. Cambridge, MA: Harvard University, John F. Kennedy School of Government.

Colburn, David R., and Lance Dehaven-Smith. 1999. *Government in the Sunshine State: Florida Since Statehood*. Gainesville: University Press of Florida.

Coleman, James. 1990. *Foundations of Social Theory*. Cambridge: Harvard University Press.

Committee on Restoration of the Greater Everglades Ecosystem (CROGEE). 2001. *Aquifer Storage and Recovery in the Comprehensive Everglades Restoration Plan: A Critique of the Pilot Projects and Related Plans for ASR in the Lake Okeechobee and Western Hillsboro Areas*. Washington, DC: National Academies Press. http://books.nap.edu/books/0309073472/html/26.html (accessed February, 2003).

Connelly, James, and Graham Smith. 1999. *Politics and the Environment–From Theory to Practice*. New York: Routledge.

Consensus Building Institute. 1999. *Study on the Mediation of Land Use Disputes*. Boston: Lincoln Institute of Land Policy.

Cooke, G. Dennis, Eugene B. Welch, Spencer A. Peterson, and Peter R. Newroth. 1993. *Restoration and Management of Lakes and Reservoirs*. Boca Raton, FL: Lewis Publishers.

Crawford, G.B. 2002. Farm Producers Earn Recognition for Their Environmental Stewardship. *FloridAgriculture* (August). http://floridafarmbureau.org/flag/aug2k2/cspread.html (accessed February 16, 2004).

Cronin, John, and Robert F. Kennedy, Jr. 1997. *The Riverkeepers: Two Activists Fight to Reclaim our Environment as a Basic Human Right*. New York: Scribners.

Davies, J. Clarence, and Jan Mazurek. 1998. *Pollution Control in the United States: Evaluating the System*. Washington, DC: Resources for the Future.

Dean, Henry. 2001. Letter from Henry Dean, Executive Director of SFWMD to Shannon Estenez, WWF South Florida Everglades Representative, November 6.

DeGrove, John M. 1984. *Land, Growth, and Politics*. Chicago: Planners Press.

———. 1999. Restoring Balance to Florida's System of Growth Management and Environmental Protection. In *Charting Florida's Future*, edited by Lance deHaven-Smith and Dena Hurst. Tallahassee: Florida Institute of Government, Florida State University.

Deloria, Vine, Jr. 1997. *Red Earth, White Lies: Native Americans and the Myth of Scientific Fact*. Golden, CO: Fulcrum Publishing.

Deuerling, Richard. 2003. Personal communication from Richard Deuerling, FDEP UIC program manager, to Eberhard Roeder, March 20.

Doak, Joe. 1998. Changing the World Through Participative Action: The Dynamics and Potential of Local Agenda 21. In *Participation and the Quality of Environmental Decision Making,* edited by Frans H.J.M. Coenen, Dave Huitema, and Laurence J. O'Toole, Jr. Boston: Kluwer Academic Publishers.

Duckworth, Erika N. 1996. Water Activists Describe Pinellas Lawsuit as Intimidation. *St. Petersburg Times,* May 5.

Dukes, Franklin E., and Karen Firehock. 2001. *Collaboration: A Guide for Environmental Advocates*. Charlottesville: University of Virginia.

Dzurik, Andrew. 2002. *Water Resources Planning.* 3rd ed. New York: Rowman and Littlefield.

Easton, David. 1965. *A Framework for Political Analysis.* Englewood Cliffs, NJ: Prentice-Hall.

The Economist. 1998. Water, Water Everywhere. *The Economist* (February 21)346: 27.

Edwards, Chuck. 2003. Speech by Chuck Edwards, poultry farmer, at the American Farm Bureau Federation National Environmental Issues Round Table, Suwannee River Partnership Demonstration Farm Site Visit, May 29.

Ellison, Don. 2003. Personal communication from Don Ellison, SWFWMD ASR project manager, to Eberhard Roeder, March 28.

Environmental Law Institute. 1984. *Citizen Suits: An Analysis of Citizen Enforcement Actions under EPA-Administered Statutes.* Washington, DC: Environmental Law Institute.

———. 2003. *Homeland Security and Drinking Water: An Opportunity for Comprehensive Protection of a Vital Natural Resource.* Washington, DC: Environmental Law Institute.

Erhmann, John R., and Barbara L. Stinson. 1999. Joint Fact-Finding and the Use of Technical Experts. In *The Consensus Building Handbook: A Comprehensive Guide to Reaching Agreement,* edited by Lawrence Susskind, Sarah McKearnan, and Jennifer Thomas-Larmer. Thousand Oaks, CA: Sage Publications.

Fernald, Edward, and Elizabeth D. Purdum. 1998. *Water Resources Atlas of Florida.* Tallahassee: Institute of Science and Public Affairs, Florida State University.

Fineout, Gary. 2003. Rodman Reservoir Measure Approved. *Gainesville Sun*, May 3.

Finkel, Steve E., and Edward N. Muller. 1998. Rational Choice and the Dynamics of Collective Political Action: Evaluating Alternative Models with Panel Data. *American Political Science Review* 92: 37–49.

Finkel, Steve. E., Edward N. Muller, and Karl-Dieter Opp. 1989. Personal Influence, Collective Rationality, and Mass Political Action. *American Political Science Review* 83: 885–903.

Fischer, Frank. 2000. *Citizens, Experts, and the Environment: The Politics of Local Knowledge.* Durham, NC: Duke University Press.

Fisher, Roger, and William Ury. 1983. *Getting to Yes.* New York: Viking Penguin.

Fisher, Roger, William Ury, and Bruce Patton. 1991. *Getting to Yes: Negotiating Agreement Without Giving In*, 2nd ed. New York: Penguin Books.

FloridAgriculture Viewpoint. 2000. Progress on the Suwannee. November. http://floridafarmbureau.org/flag/nov2k/viewnov.html (accessed February 16, 2004).

Florida-Agriculture.com. 1999. Poultry Producers Taking Steps to Improve Management Practices. Agriculture Press Release. July 12. http://www.florida-agriculture.com/news/071299.htm (accessed February 16, 2004).

Florida Conflict Resolution Consortium (FCRC). 2002a. East-Central Florida Water Supply Planning Initiative. Round #1 Subregional Workshops. Executive Summary. http:// consensus.fsu.edu/ECWS/index.html (accessed April 20, 2003)

———. 2002b. East-Central Florida Water Supply Planning Initiative. Round #2 Subregional Workshops. Executive Summary. http://consensus.fsu.edu/ECWS/index.html (accessed April 20, 2003).

———. 2002c. East-Central Florida Water Supply Planning Initiative. Round #3 Subregional Workshops. Executive Summary. http://consensus.fsu.edu/ECWS/index.html (accessed April 20, 2003).

———. 2002d. East-Central Florida Water Agenda: A Report on the Water Supply Planning Initiative Process. November 2002. http://consensus.fsu.edu/ECWS/index.html (accessed April 17, 2003).

———. 2002e. Proceedings of the East-Central Florida Water Summit. January 31, 2002. http://consensus.fsu.edu/ECWS (accessed April 21, 2003).

Florida Council of One Hundred. 2003. *Improving Florida's Water Supply Management Structure: Ensuring and Sustaining Environmentally Sound Water Supplies and Resources to Meet Current and Future Needs.* Tampa: Florida Council of One Hundred.

Florida Defenders of the Environment. 2003a. Chronology of Significant Events in the Lower Ocklawaha River Valley. http://www.fladefenders.org/publications/chronology.html (accessed January 18, 2003).

———. 2003b. Restoring the Ocklawaha—The Barge Canal. http://www.fladefenders.org/publications/restoring3.html (accessed January 18, 2003).

Florida Department of Agriculture and Consumer Services (FDACS). 2003. Florida Agricultural Fast Facts. http://www.florida-agriculture.com (accessed February 26, 2004).

Florida Department of Environmental Protection (FDEP). 1995. *Delineation of Ground and Surface Water Area Potential Impacted by an Industrial Discharge to the Fenholloway River of Taylor County, Florida.* Tallahassee: Division of Water Facilities.

———. 2001. "Memorandum of Understanding Between the Florida Department of Environmental Protection and the Clean Water Network/Natural Resources Defense Council and Buckeye Florida L.P. for The Fenholloway River Evaluation Initiative." Dated December 17, 2001.

———. 2003. Surface Water: Outstanding Florida Waters (April 10). http://www.dep.state.fl.us/water/surfacewater/ofw.htm (accessed February 16, 2004).

———. No date a. Middle Suwannee River Hydrologic Unit Area Animal Waste Disposal Best Management Practices Demonstration Project. http://www.dep.state.fl.us/water/nonpoint/docs/319h/WM482clo.pdf (accessed February 12, 2004).

———. No date b. More Protection, Less Process: Common-Sense Environmental Management. http://www.dep.state.fl.us/secretary/info/pubs/brochure.pdf (accessed February 16, 2004).

Florida Public Service Commission. 2001. *Water Allocation Markets.* Tallahassee: Florida Public Service Commission, Division of Policy and Intergovernmental Liaison.

Florida Senate. 2001. Information about Senate Bill 0854 and House Bill 0705. www.flsenate.gov/data/session/2001/ (accessed February through May 2003).

Flyvbjerg, Bent. 1999. *Rationality and Power: Democracy in Practice.* Chicago: University of Chicago Press.

Forester, John. 1994. Lawrence Susskind: Activist Mediation and Public Disputes. In *When Talk Works: Profiles of Mediators*, edited by Deborah M. Kolb and Associates. San Francisco: Jossey-Bass.

———. 1999. *The Deliberative Practitioner: Encouraging Participatory Planning Processes.* Cambridge: MIT Press.

———. 2004a. *Dispute Resolution Meets Policy Analysis, Or Native Gathering Rights on "Private" Lands? A Profile of Peter Adler.* Ithaca, NY: Cornell University, Department of City and Regional Planning.

———. 2004b. Planning and Mediation, Participation and Posturing: What's a Deliberative Practitioner to Do? Paper prepared for the Annual Symposium of the Interdisciplinary Ph.D. Program in Urban Design and Planning University of Washington, Seattle, April 15.

———. 2005. *The Drama of Mediation* (draft prepared for publication, available via Department of City and Regional Planning, Cornell University, Ithaca NY).

Forester, John, and I. Weiser (eds.). 1996. *Making Mediation Work: Profiles of Environmental and Community Mediators.* Typescript. Department of City and Regional Planning. Ithaca, NY: Cornell University.

Freeze, R.A., and J.A. Cherry. 1979. *Groundwater.* Englewood Cliffs, NJ: Prentice Hall.

Friedman, Robert. 1994. Boiling Down the Regional Water Crisis: It's Really About the Flow of Money. *St. Petersburg Times*, July 3.

Friedmann, John. 1987. *Planning in the Public Domain: From Action to Knowledge.* Princeton, NJ: Princeton University Press.

Fudenberg, Drew, and Jean Tirole. 1992. *Game Theory.* Cambridge: MIT Press.

Galantowicz, Richard E., and John R. Shuman. 1994. Volume I, Executive Summary. In *Envi-*

ronmental Studies Concerning Four Alternatives for Rodman Reservoir and the Lower Ocklawaha River. Palatka, FL: St. Johns River Water Management District.

Garcia, Wayne. 1993. Water Pipeline Clogged with Uncertainty. *St Petersburg Times*, May 14.

Garcia, Wayne, and David K. Rogers. 1993. Pinellas Says No to Proposed Watering Ban. *St. Petersburg Times*, August 18.

Glasgow, T. Nigel. 1999. Suwannee River Basin Nutrient Management Partnership: A Voluntary Incentive-Based Effort to Maintain Agricultural Production and Protect Water Quality in Florida. Abstract from Wisconsin Academy of Sciences, Arts and Letters Building on Leopold's Legacy: Conservation for a New Century Conference, Madison, WI, October 4–7. http://www.wisconsinacademy.org/landethic/wa2.htm (accessed February 16, 2004).

Glazer, Amihai, and Lawrence S. Rothenberg. 2001. *Why Government Succeeds and Why it Fails.* Cambridge: Harvard University Press.

Glenn, John S. 2003. Safe Drinking Water. *Pelican* 35(1): 12.

Glennon, Robert. 2002. *Water Follies: Groundwater Pumping and the Fate of America's Fresh Waters.* Washington, DC: Island Press.

Governor's Commission for a Sustainable South Florida. 1995. *Initial Report,* October 1. Coral Gables, FL.

Graf, William L., ed. 2002. *Dam Removal Research: Status and Prospects.* Washington, DC: Heinz Center.

Greenwood, Davydd, and Morten Levin. 1999. *Introduction to Action Research: Social Research for Social Change.* Thousand Oaks, CA: Sage.

Grumbine, R. Edward. 1994. What is Ecosystem Management? *Conservation Biology* 8(1): 27–38.

Gunderson, Lance, C.S. Holling, and S. Light (eds.). 1995. *Barriers and Bridges to the Renewal of Ecosystems and Institutions.* New York: Columbia University Press.

Hahn, Robert W. 1990. The Political Economy of Environmental Regulation: Toward a Unifying Theory. *Public Choice* 110: 353–377.

Hall, Frankie. 2003. Personal communication between Frankie Hall, Assistant Director of Agricultural Policy at the Florida Farm Bureau Federation, and the author. January 29.

Hall, Peter A. 1993. Policy Paradigms, Social Learning, and the State: The Case of Economic Policymaking in Britain. *Comparative Politics* 23(3): 275–296.

Hardin, Russell. 1982. *Collective Action.* Baltimore, MD: Johns Hopkins University Press.

Harter, Philip. 1992. Negotiated Regulations: A Cure for Malaise. *Georgetown Law Journal* 71: 1–118.

Harvey, Richard. 2003. Personal communication from Richard Harvey, USEPA Region IV Miami field office director, to Eberhard Roeder. April 28.

Hauserman, Julie. 1991. Florida's Forgotten River. *Tallahassee Democrat*, March 17, 6A.

Healey, Patsy. 1997. *Collaborative Planning: Shaping Places in Fragmented Societies.* Vancouver: University of British Columbia Press.

Heclo, Hugh. 1974. *Social Policy in Britain and Sweden.* New Haven: Yale University Press.

Heifetz, Ronald A. 1994. *Leadership without Easy Answers.* Cambridge, MA: Belknap Press.

Heifetz, Ronald A., and Donald Laurie. 1997. The Work of Leadership. *Harvard Business Review* 124 (January–February).

Heller, Jean. 2001. Arbitrators Clear Way for Reservoir Construction. *St. Petersburg Times,* May 16.

Hemphill, Rod. 2000. "Farm Bureau's CARES Program to Recognize Good Environmental Practices." News Release. Florida Farm Bureau Federation. http://floridafarmbureau.org/flag/jan00/suwan.html (accessed March 2003).

Hillier, Jean. 2002. *Shadows of Power: An Allegory of Prudence in Land-use Planning.* London: Routledge.

Hock, Dee. 1999. *The Birth of the Chaordic Age.* Los Angeles: Pub Group West.

Hodges, A.W., and W.D. Mulkey. 2003. Regional Economic Impacts of Florida's Agricultural and Natural Resource Industries. http://economicimpact.ifas.ufl.edu/publications/Fla_Ag_Nat_Resource_Ind.pdf (accessed February 29, 2004).

Holling, C.S. 1978. *Adaptive Environmental Assessment and Management.* New York: John Wiley & Sons.

Hollingsworth, Jan. 2001. EPA Tests Uncover Dioxin Pollution. *The Tampa Tribute*, February 9, 1, 10.

Holmbeck-Pelham, Skelly. 2003. Personal communication to Steve Leitman from Skelly Holmbeck-Pelham, Coordinator, Tri-state Conservation Coalition, Atlanta. December.

Horkan, Jacquelyn. May/June 1999. "Permission Denied: The Trouble with Pipelines,"in Florida Business Insight, http://flabusinessinsight.com/1999Issues/may&june99/may-covertext.htm (accessed January 12, 2003).

Hornsby, David, and Rob Mattson. 1998. *Surfacewater Quality and Biological Monitoring Network Annual Report, 1997.* WR-98-03. Live Oak, FL: SRWMD.

Hornsby, David, Rob Mattson, and Tom Mirti. 2002a. *Surface Water Quality and Biological Annual Report 2001.* WR-01-02-04. Live Oak, FL: SRWMD. http://www.srwmd.state.fl.us/resources/surfacewater+quality+and+biological+annual+report+.pdf (accessed February 16, 2004).

Hornsby, David, Ron Ceryak, and Warren Zwanka. 2002b. *Groundwater Quality Report 2001.* WR-01-02-05. Live Oak, FL: SRWMD.

Horvath, Glenn. 2003. Personal communication between Glenn Horvath, SRWMD, and the author. January 29.

Howell, W. Mike, A. Ann Black, and Stephen A. Bortone. 1980. Abnormal Expression of Secondary Sex Characters in a Population of Mosquitofish, Gambusia Affinis holbrooki: Evidence for environmentally-induced masculinization. *COPEIA* 4: 676–681. http://www.ejnet.org/rachel/rehw475.htm#2 (accessed April 23, 2003).

Hui, Anthony M. 1999. Archaic Rules on Deep Wells Need Reform. *Enviro-Net* August. http://www.enviro-net.com (accessed February 14, 2004)

Hull, Victor. 2001a. Water-storage Legislation Appears Doomed. *The (Lakeland) Ledger,* May 1. http://web.lexis-nexis.com/universe (accessed February 2003).

________. 2001b. Activists Discuss Amendment as Strategy in Aquifer Issue. *Sarasota Herald-Tribune,* June 14. http://web.lexis-nexis.com/universe (accessed February 2003).

Hurwitz, Jon, and Mark Peffley. 1987. How Are Foreign Policy Attitudes Structured? A Hierarchical Model. *American Political Science Review* 81: 1099–1120.

Ingram, Helen. 1990. *Water Politics: Continuity and Change.* Albuquerque: University of New Mexico Press.

Innes, Judith, and David Booher. 1999. Consensus Building and Complex Adaptive Systems. *Journal of the American Planning Association* 65: 412–423.

Institute of Food and Agricultural Services (IFAS). 2002. TMDLs and the Suwannee River Partnership. *Poop Scoop* 2. http://nfrec-sv.ifas.ufl.edu/2002-02_poopscoop.htm#TMDLs%20and%20the (accessed July 1, 2003).

Jacobs, Robert T. 2002. Official USDA (Appeal Deciding Officer, Regional Forester) Correspondence Containing Final Administrative Determination of the Status of Appeal 0-08-0020 of Forest Supervisor Marsha Kearney's December 17, 2001 Decision for the Occupancy and Use of National Forest Lands and Ocklawaha River Restoration on the Ocala National Forest in Florida. April 5.

Jasanoff, Sheila. 1991. American Exceptionalism and the Political Acknowledgement of Risk. *Daedalus* 119: 61–81.

Jehl, Douglas. 2001. Florida, Low on Drinking Water, Asks E.P.A. to Waive Safety Rule. *The New York Times,* April 13. http://www.voteenvironment.org/statenews/FL041301.htm (accessed February 07, 2003).

John, DeWitt. 1994. *Civic Environmentalism.* Washington: Congressional Quarterly Press.

Johnson, Neil. 2003. EPA may ease its drinking water rules. *Tampa Tribune,* Jun 26. http://www.tampatrib.com/FloridaMetro/MGASK849EHD.html (accessed July 14, 2003).

Joiner, Jerry. 2003. Speech of Jerry Joiner, USDA NRCS, at Suwannee River Partnership Steering Committee Meeting, May 28.

Jones, Curry. 2003. Speech of Curry Jones, EPA Region 4, TMDL Section, at Suwannee River Partnership Steering Committee Meeting, May 28.

Jones, Robert. 1996a. Building Consensus on Florida's Water Resources. Excerpts from *Solutions Newsletter* February 6. http://consensus.fsu.edu/solutions/sol2963.html (accessed April 2, 2004).

———. 1996b. North Tampa Bay Joint Science Review Process Final Report. Tallahassee: Florida Conflict Resolution Consortium.

———. 1997. Building an Intergovernmental Coordination Model in the Everglades Ecosystem. *National Institute for Dispute Resolution NEWS* IV: 5.

———. 2002. The Everglades Mediation: Reframing the Politics of Consensus. In *Finding the Common Ground*, ed. Peter S. Adler. Manuscript.

———. 2003a. Personal communication from Robert Jones, Director FCRC and member of the Facilitation Team, March 3.

———. 2003b. Personal communication between Robert M. Jones, Director, Florida Conflict Resolution Consortium, and the author. June 2.

Jones, Robert M., et al. 1992. *Report of the Committee on Qualifications.* Washington, DC: Society for Professionals in Dispute Resolution.

Joyner, Daryll. 2003. Speech of Daryll Joyner, TMDL program administrator, FDEP Bureau of Watershed Management, at Suwannee River Partnership Steering Committee Meeting, January 29.

Kagan, Robert A. 2001. *Adversarial Legalism: The American Way of Law.* Cambridge: Harvard University Press.

Kahn, Matthew E., and John G. Matsusaka. 1997. Demand for Environmental Goods: Evidence from Voting Patterns on California Initiatives. *Journal of Law and Economics* 40: 137–173.

Kanter, Rosabeth Moss. 1999. The Enduring Skills of Change Leaders. *Leader to Leader* 13(Summer): 15–22.

Kam, Dara. 2001. House Considers Experimental Water Storage. *Associated Press* 04/10/2001. http://web.lexis-nexis.com/universe (accessed February 2003).

Kasperson, Roger E., Dominic Golding, and Seth Tuler. 1992. Social Distrust as a Factor in Siting Hazardous Facilities and Communicating Risks. *Journal of Social Issues* 4:161-187.

Katz, Brian G., H. David Hornsby, Johnkarl F. Bohlke, and Michael F. Mokray. 1999. *Sources and Chronology of Nitrate Contamination in Spring Waters, Suwannee River Basin, Florida.* Water-Resources Investigations Report 99-4252. Prepared in cooperation with the Suwannee River Water Management District. Tallahassee, FL: U.S. Geological Survey.

Katz, Brian G., and H. David Hornsby. 1998. *A Preliminary Assessment of Sources of Nitrate in Springwaters, Suwannee River Basin, Florida.* Open-File Report 98-69. Prepared in cooperation with the Suwannee River Water Management District. Tallahassee, FL: U.S. Geological Survey. http://fl.water.usgs.gov/PDF_files/ofr98_69_katz.pdf (accessed February 16, 2004).

Kearney, Marsha. 2002. Official USDA/FS, National Forests in Florida (Forest Supervisor) Correspondence with Bob Ballard, Deputy Secretary, Land and Recreation, DEP Regarding Special Use Permit for Occupation of National Forest Lands. July 26.

Kemper, Carol. 2003. Speech of Carol Kemper, EPA Region 4, at American Farm Bureau Federation National Environmental Issues Round Table, Suwannee River Partnership Demonstration Farm Site Visit, May 29.

Kenney, D.S. 1996. Review of Coordination Mechanisms with Water Allocation Responsibilities. Paper prepared for the ACT-ACF Comprehensive Study. Mobile, AL: US Army Corps of Engineers, Mobile District.

Keohane, Nathaniel O., Richard L. Revesz, and Robert N. Stavins. 1999. The Positive Political Economy of Instrument Choice in Environmental Policy. In *Environmental and Public Economics: Essays in Honor of Wallace Oates,* edited by A. Panagariya, P. R. Portney, and R.M. Schwab. London: Edward Elgar.

Kirchgässner, Gebhard, and Friedrich Schneider. 2003. On the Political Economy of Environmental Policy. *Public Choice* 115: 369–396.

Krizek, K., and J. Power. 1996. *A Planners Guide to Sustainable Development.* Planning Advisory Service Report No. 467. Chicago: American Planning Association.

Kuhn, Thomas S. 1962. *The Structure of Scientific Revolutions.* Chicago: University of Chicago.

Kwiatkowski, Peter. 2003. Personal communication from Peter Kwiatkowski, SFWMD ASR project manager, to Eberhard Roeder. April 2 and May 15.

Ladd, Everett Carl, and Karlyn H. Bowman. 1995. *Attitudes Toward the Environment: Twenty-five*

Years after Earth Day. Washington, DC: American Enterprise Institute.

Langton, Stuart. 1984. *Environmental Leadership.* Lexington, MA: Lexington Books.

Langton, Stuart, and Terrance Salt. 2003. Working Together: Collaboration. In *Adaptive Governance and Florida's Water Conflicts: The Case Studies*, edited by Aysin Dedekorkut, John Scholz, and Bruce Stiftel. Tallahassee, FL: Florida State University, DeVoe L. Moore and Family Center for the Study of Economic Policy and Government, June, 243-254.

Latour, Bruno. 1979. *Life in the Laboratory.* Beverly Hills, CA: Sage Publications.

Lax, David, and James Sebenius. 1987. *The Manager as Negotiator.* New York: Free Press.

Leach, William. 2002. Surveying Diverse Stakeholder Groups. *Society and Natural Resources* 15: 641–649.

Leach, William, and Neil Pelkey. 2001. Making Watershed Partnerships Work: A Review of the Empirical Literature. *Journal of Water Resources Planning and Management* 127 (November/December): 378–385.

Leach, William, and Paul Sabatier. 2003. Facilitators, Coordinators, and Outcomes. In *The Promise and Performance of Environmental Conflict Resolution*, edited by Rosemary O'Leary and Lisa Bingham. Washington, DC: Resources for the Future.

———. 2005. Are Trust and Social Capital the Keys to Success? In *Swimming Upstream: Collaborative Approaches to Watershed Management*, edited by Paul Sabatier, Will Focht, Mark Lubell, Zev Trachtenberg, Arnold Vedlitz and Marty Matlock. Cambridge: MIT Press.

Leadership Florida. 2003. http://www.leadershipflorida.org (accessed February 20, 2004).

Lee, Kai N. 1993. *Compass and Gyroscope: Integrating Science and Politics for the Environment.* Washington, DC: Island Press.

Legal Environmental Assistance Foundation (LEAF). 2001. Aquifer Storage and Recovery: What's at Stake? http://www.orckl.com/dwi2/jointposition.htm (accessed May 20, 2003).

Leitman, S.F., J. Dowd, and S. Holmbeck-Pelham. 2003. An Evaluation of Observed and Unimpaired Flows and Precipitation during Three Drought Events in the ACF Basin. In Proceedings of the 2003 Georgia Water Resources Conference.

Lessard, G. 1998. An Adaptive Approach to Planning and Decision-Making. *Landscape and Urban Planning* 40: 81–87.

Light, Stephen S., and J. Walter Dineen. 1994. Water Control in the Everglades: A Historical Perspective. In *Everglades: The Ecosystem and Its Restoration*, edited by Steven M. Davis and John C. Ogden. Delray Beach, FL: St. Lucie Press.

Light, Stephen S., Lance H. Gunderson, and C.S. Holling. 1995. The Everglades: Evolution of Management in a Turbulent Ecosystem. In *Barriers and Bridges to the Renewal of Ecosystems and Institutions*, edited by Stephen S. Light, Lance H. Gunderson, and C.S. Holling. New York: Columbia University Press.

Lijphart, Arend. 1999. *Patterns of Democracy: Government Forms and Performance in Thirty-Six Countries.* New Haven: Yale University Press.

Lindblom, Charles, and D. Cohen. 1979. *Usable Knowledge: Social Science and Social Problem Solving.* New Haven: Yale University Press.

Loop, Carl B., Jr. 2001. Program Shows that Farmers Care. *FloridAgriculture.* August. http://www.fb.com/flfb/flag/aug2k1/presaug.html (accessed July 1, 2003).

Lord, Charles, Lee Ross, and Mark Lepper. 1979. Biased Assimilation and Attitude Polarization: The Effects of Prior Theories on Subsequently Considered Evidence. *Journal of Personality and Social Psychology* 37: 2098–2109.

Lowry, William R. 2003. *Dam Politics: Restoring America's Rivers.* Washington, DC: Georgetown University Press.

Lubell, Mark. 2004. Collaborative Watershed Management: A View from the Grassroots. *Policy Studies Journal* 32(3): 321-341.

———. 2003. Collaborative Institutions, Belief-systems, and Perceived Policy Effectiveness. *Political Research Quarterly* 56(3): 309–323.

———. 2002a. Environmental Activism as Collective Action. *Environment and Behavior* 34: 431–454.

———. 2002b. Collaborative Environmental Institutions: All Talk and No Action? *Journal of Policy Analysis and Management* 23(3):549–573.

Maass, Arthur, and Raymond L. Anderson. 1978. *And the Desert Shall Rejoice: Conflict, Growth and Justice in Arid Environments*. Cambridge: MIT Press.

Maloney, Frank E., Richard C. Ausness, and J. Scott Morris. 1972. *Model Water Code, With Commentary*. Gainesville: University of Florida Press.

Mann, Doug. 2001. *Utility Council Report 2000–2001,* Florida Section of American Water Works Association, Legislative Session Final Report: 4–10. www.fsawwa.org/utilreport01.pdf (accessed May 20, 2003).

Mansfield, Geofrey. 2003. Personal communication from Geofrey Mansfield, FDEP legislative liaison, to Eberhard Roeder. May 28.

Marcus, Alfred A., Donald A. Geffen, and Ken Sexton. 2002. *Reinventing Environmental Regulation: Lessons from Project XL*. Washington, DC: Resources for the Future.

Marella. R.L. 1999. Water Withdrawals, Use, Discharge, and Trends in Florida, 1995. Water-Resources Investigations Report 99-4002. Washington, DC: U.S. Geological Survey.

Marshall, Gary, and Connie P. Ozawa. 2004. Mediation at the Local Level: Implications for Democratic Governance. In *Tampering with Tradition: The Unrealized Authority of Democratic Agency,* edited by Peter Bogason, Sandra Kensen, and Hugh Miller. Lanham, MD: Lexington Books.

Maser, Chris. 1997. *Sustainable Community Development: Principles and Concepts.* Delray Beach, FL: St. Lucie Press.

Matthews, Frank E., and Kimberly A. Grippa. 1997. Governor's Water Resources Coordinating Council Hears from WMDs, DEP. *Florida Environmental Compliance Update* December 8(11).

McCarty, Nolan, and Lawrence S. Rothenberg. 1996. Commitment and the Campaign Contribution Contract. *American Journal of Political Science* 40: 872–904.

McClurg, Sue. 2002. *The Water Forum Agreement: A Model for Collaborative Problem Solving.* Sacramento, CA: Water Education Foundation.

McConnell, Grant. 1966. *Private Power and American Democracy.* New York: Vintage Press.

McDonald, Geoffrey T. 1996. Planning as Sustainable Development. *Journal of Planning Education and Research* 15(3): 225–236.

McFarland, Andrew S. 1993. *Cooperative Pluralism: The National Coal Policy Experiment.* Lawrence: University of Kansas Press.

McGrail, Linda, Kenneth Berk, Donald Brandes, Douglas Munch, Clifford Neubauer, William Osburn, Donthamsetti Rao, John Thomson, and David Toth. 1998. St. Johns River Water Management District. In *Water Resources Atlas of Florida*, edited by Edward A. Fernald and Elizabeth D. Purdum. Tallahassee: Institute of Science and Public Affairs, Florida State University.

Meinhart, David. W. 1989. Promoting Regional Water Supply Development: A Case Study from Southwest Florida. In *Water: Laws and Management,* edited by Frederick E. Davis. Middleburg, VA: American Water Resources Association.

Membrane and Separation Technology News. 2005. March: 23 (6).

Merritt, M.L. 1985. *Subsurface Storage of Freshwater in South Florida: A Digital Model Analysis of Recoverability*. Prepared in Cooperation with the U.S. Army Corps of Engineers. USGS Water-Supply Paper 2261. Alexandria, VA: US Government Printing Office.

Metcalf, John G. 2002. Submission of Notice of Appeal on Behalf of Save Rodman Reservoir, Inc. with regard to Forest Service's Record of Decision for the Occupancy and Use of National Forest Service Lands and Ocklawaha River Restoration, Ocala National Forest. February 21.

Meyer, Stephen M. 2001. Community Politics and Endangered Species Protection. In *Protecting Endangered Species in the United States,* edited by Jason F. Shogren and John Tschirhart. New York: Cambridge University Press.

Miami Herald. 2001a. Groups 'Miles Apart' on Plan to Store Excess Polluted Water. April 28. http://exchange.law.miami.edu/everglades/news/2001/ (accessed February 14, 2004).

———. 2001b. Water Law Fell Victim to Hysteria. Editorial. May 02. http://exchange.law.miami.edu/everglades/news/2001/ (accessed February 14, 2004).

Milbrath, Lester, and M.L. Goal. 1977. *Political Participation*. 2nd ed. Chicago: Rand McNally.

Mill J.S. 1962. On the Connection between Justice and Utility. In *Utilitarianism, On Liberty and Essay on Bentham*, edited by Mary Warnock. Fontana Press, 296-321.

Missimer, Thomas, 2003. Summary of Research on Agricultural Water Use in Florida with Comparison to Natural System Water Use. Report to the Florida Department of Agriculture and Consumer Affairs. Tallahassee: Florida Department of Agriculture and Consumer Affairs.

Missimer, Thomas M., Weixing Guo, Charles W. Walker, and Robert G. Maliva. 2002. Hydraulic and Density Considerations in the Design of Aquifer Storage and Recovery Systems. *Florida Water Resources Journal* February: 30–36.

Moe, Terry M. 1980. *The Organization of Interests*. Chicago: University of Chicago Press.

Moe, Terry M., and Michael Caldwell. 1994. The Institutional Foundations of Democratic Governments—A Comparison of Presidential and Parliamentary Systems. *Journal of Institutional and Theoretical Economics* 150: 171–195.

Moncada, Carlos. 1995a. Water War Turns to Words: County Commissioners Want Counterpunch to Public Messages Being Sent in Ongoing Water Crisis. *St. Petersburg Times,* October 11.

———. 1995b. Latvala Dives into Middle of Water Wars. *The Tampa Tribune,* June 15.

Mueller, D.C. 1976. Public Choice: A Survey. *Journal of Economic Literature* 14: 395–433.

Multinational Monitor. 1991. Procter & Gamble. 12(12). http://multinationalmonitor.org/hyper/issues/1991/12/mm1291_07.html (accessed January 12, 2003).

Napier, Ted L., and Silvana M. Camboni. 1988a. Attitudes Toward a Proposed Soil Conservation Program. *Journal of Soil and Water Conservation* 43: 186–191.

Napier, Ted L., Cameron S. Thraen, and Silvana M. Camboni. 1988b. Willingness of Land Operators to Participate in Government-sponsored Soil Erosion Control Programs. *Journal of Rural Studies* 4: 339–347.

Napier, Ted L., Cameron S. Thraen, and Stephen L. McClaskie. 1988c. Adoption of Soil Conservation Practices by Farmers in Erosion-Prone Areas of Ohio: The Application of Logit Modeling. *Society and Natural Resources* 1: 109–129.

Napier, Ted L., and Mark Tucker. 2001. Use of Soil and Water Protection Practices Among Farmers in Three Midwest Watersheds. *Environmental Management* 27: 269–279.

National Performance Review. 1993. *Environmental Protection Agency: Accompanying Report of the National Performance Review.* Washington, DC: Office of the Vice President.

National Research Council Committee on Restoration of Aquatic Ecosystems (NRC-CRAE). 1992. *Restoration of Aquatic Ecosystems: Science, Technology and Public Policy*. Washington, DC: National Academy of Sciences.

Neidrauer, Calvin J., and Richard M. Cooper. 1989. *A Two-Year Field Test of the Rainfall Plan: A Management Plan for Water Deliveries to Everglades National Park.* Technical Publication 89-3 (November). West Palm Beach, FL: South Florida Water Management District.

Nelkin, Dorothy (ed.). 1992. *Controversy: The Politics of Technical Decisions.* 3rd ed. Thousand Oaks, CA: Sage Publications.

Norton, B.G., and A.C. Steinemann. 2001. Environmental Values and Adaptive Management. *Environmental Values* 10: 473–506.

Office of Agricultural Water Policy (OAWP), Florida Department of Agriculture and Consumer Services. 2004. *Agricultural Water Policy.* www.floridaagwaterpolicy.com (accessed February 28, 2004).

Olinger, David. 1994. Who Controls Our Water? *St. Petersburg Times,* June 19.

Olson, Mancur. 1965. *The Logic of Collective Action*. Cambridge: Harvard University Press.

Ostrom, Elinor. 1990. *Governing the Commons, The Evolution of Institutions for Collective Action.* Cambridge: Cambridge University Press.

———. 1992. The Rudiments of a Theory of the Origins, Survival, and Performance of Common-Property Institutions. In *Making the Commons Work: Theory, Practice, and Policy,* edited by Daniel W. Bromley. San Francisco: ICS Press.

———. 1998. A Behavioral Approach to the Rational Choice Theory of Collective Action. *American Political Science Review* 92: 1–22.

———. 1999. An Assessment of the Institutional Analysis and Development Framework. In *Theories of the Policy Process,* edited by Paul Sabatier. Boulder, CO: Westview Press.

Ostrom, Elinor, Roy Gardner, and James Walker. 1994. *Rules, Games, and Common-Pool Resources.* Ann Arbor: University of Michigan Press.

Ozawa, Connie P. 1991. *Recasting Science: Consensual Procedures in Public Policy Making.* Boulder, CO: Westview Press.

———. 1993. Improving Public Participation in Enviornmental Decisionmaking: The Use of Transformative Mediation Techniques. *Environment and Planning C: Government and Policy* 1: 103–117.

Paben, J. Zokovitch. 2003. Personal communication from J. Zokovitch Paben, LEAF Staff Attorney, to Eberhard Roeder. April 25.

Palmer, Kurt, Wallace E. Oates, and Paul Portney. 1995. Tightening Environmental Standards: The Benefit–Cost or the No-Cost Paradigm. Journal of Economic *Perspectives* 9: 119–132.

Pateman, Carole. 1970. *Participation and Democratic Theory*. New York: Cambridge University Press.

Pelham, Thomas G. 1979. *State Land-use Planning and Regulation*. Lexington, MA: Lexington Books.

Pettigrew, Richard A., and Janice Fleischer. 2002. Quality and Implementation of Agreements: Are Products of Environmental Public Policy Conflict Resolution and Consensus Building Efforts Being Implemented? How Do They Compare to Products of Traditional Forums. In *Critical Issues Papers,* ed. Association for Conflict Resolution, Environmental Public Policy Section. Washington, DC. (http://www.mediate.com/acrepp/pg5.cfm) (accessed November 15, 2004).

Pfankuch, Thomas. 2002. Florida's Rotten River. *Florida Times-Union,* June 5, A1, A7.

Pickens, Joe, and Dennis Baxley. 2003. House Bill 697, Establishment of Rodman Reservoir State Reserve.

Pilla, Jen. 1996. Pinellas to Ponder Water Supplier's Demise. *St. Petersburg Times,* July 19.

Policy Consensus Initiative. 2004. Governing Tools for the 21st Century: How State Leaders are Using Collaborative Problem Solving and Dispute Resolution. http://www.policyconsensus.org/pubs/index.html (accessed 16 November 2004).

Pollick, Michael, and Chris Davis. 2002. Dead in the Water; Enron's Failure to Privatize Resources was Factor in Collapse. *Sarasota Herald-Tribune,* February 9. http://web.lexis-nexis.com/universe (accessed February 2003).

Porter, Michael E., and Claas van der Lind. 1995. Towards a New Conception of the Environment-Competitiveness Relationship. *Journal of Economic Perspectives* 9: 97–118.

Purdum, Elizabeth D. 2002. *Florida's Waters: A Water Resources Manual from Florida's Water Management Districts*. Bartow, FL: Southwest Florida Water Management District.

Putnam, Robert. 1993. *Making Democracy Work: Civic Traditions in Modern Italy*. Princeton: Princeton University Press.

Pyne, R. David G. 1995. *Groundwater Recharge and Wells: A Guide to Aquifer Storage Recovery.* Boca Raton, FL: Lewis Publishers.

———. 2002. Aquifer Storage and Recovery: The Path Ahead. *Florida Water Resources Journal* February: 19–27.

______. 2003. Personal communication from R. David G. Pyne, President of ASR Systems LLC, to Eberhard Roeder. March 20 and 24.

Quattrone, George, and Amos Tversky. 1988. Contrasting Rational and Psychological Analyses of Political Choice. *American Political Science Review* 82: 719–736.

Quirk, Paul J. 1989. The Cooperative Resolution of Policy Conflict. *American Political Science Review* 83(3): 905–921.

Raiffa, Howard. 1985. *The Art and Science of Negotiation*. Cambridge: Harvard University Press.

Rand, Honey. 2000. In the Public Interest: A Story of Conflict, Communication, and Change in Tampa Bay's Water Wars. PhD diss., University of South Florida.

———. 2003. *Water Wars: a Story of People, Politics, and Power*. Philadelphia: Xlibris.

Reese, Ronald. 2002. *Inventory and Review of Aquifer Storage and Recovery in Southern Florida*. U.S. Geological Survey Water-Resources Investigations Report 02-4036. Tallahassee, FL: U.S. Geological Survey.

Reisner, Marc. 1986. *Cadillac Desert: The American West and Its Disappearing Water*. New York: Penguin Books.

Ritchie, Bruce. 2002. Pollution Seeps into Springs. *Tallahassee Democrat,* August 11. http://www.tallahassee.com/mld/tallahassee/3839726.htm?template (accessed February 16, 2004).

Rittel, Horst, and Melvin M. Webber. 1973. Dilemmas in a General Theory of Planning. *Policy Sciences* 4: 155–169.

Roberts, Martha. 2003a. Speech of Martha Roberts, Deputy Commissioner of Agriculture, FDACS, at Suwannee River Partnership Steering Committee Meeting, April 16.

———. 2003b. Speech of Martha Roberts, Deputy Commissioner of Agriculture at American Farm Bureau Federation National Environmental Issues Round Table Suwannee River Partnership Presentation, May 29.

Roseland, Mark. 1992. *Toward Sustainable Communities*. Ottowa: National Round Table on the Environment and the Economy.

Rothenberg, Lawrence S. 2002. *Environmental Choices: Policy Responses to Green Demands.* Washington, DC: Congressional Quarterly Press.

Rubin, Jeffrey Z., and Frank E.A. Sander. 1988. When Should We Use Agents? Direct versus Representative Negotiation. *Negotiation Journal* 4: 395–401.

Ruhl, J.B. 2003. Equitable Apportionment of Ecosystem Services: New Water Law for a New Water Age. *Journal of Land Use and Environmental Law* 19(1): 47–57.

Ruhl, Suzi. 1999. Statutory Rape. *The Environmental Forum* 16(5): 19–27.

———. 2003. Personal communication from Suzi Ruhl, LEAF founder, to Eberhard Roeder. November 22.

Sabatier, Paul, and Hank Jenkins-Smith (eds.). 1993. *Policy Change and Learning: An Advocacy Coalition Approach.* Boulder, CO: Westview Press.

———. 1999. The Advocacy Coalition Framework: An Assessment. In *Theories of the Policy Process,* edited by Paul Sabatier. Boulder, CO: Westview, 117–168.

Sabatier, Paul, and Matthew Zafonte. 2001. Policy Knowledge, Advocacy Organizations. In *International Encyclopedia of the Social and Behavioral Sciences,* edited by Neil Smelser and Paul Baltes. London: Pergamon Press.

———. 2002. Are Scientists and Bureaucrats Members of Advocacy Coalitions? In *An ACF Lens on Environmental Policy*, edited by Paul Sabatier. Manuscript.

Sabatier, Paul, Susan Hunter, and Susan McLaughlin. 1987. The Devil Shift: Perceptions and Misperceptions of Opponents. *Western Political Quarterly* 40(Sept): 449–476.

Sabatier, Paul, Will Focht, Mark Lubell, Zev Trachtenberg, Arnold Vedlitz, and Marty Matlock. 2005. *Swimming Upstream: Collaborative Approaches to Watershed Management.* Cambridge: MIT Press.

Sager, Tore. 2002. *Democratic Planning and Social Choice.* Burlington, VT: Ashgate.

St. Petersburg Times. 1995. The Water Winners. June 28.

St. Petersburg Times. 2001. Editorial. Saving the Governor from Himself. May 5. http://web.-lexis-nexis.com/universe (accessed February 12, 2003).

Salamone, Debbie. 2002. Florida's Water Crisis: Science on Trial. *Orlando Sentinel,* September 22. http://www.orlandosentinel.com/news/local/state/orl-asecwater22092202sep22(0,304828).story (accessed January, 2003).

Salinero, Mike. 2001a. Aquifer Storage Passes in Senate. *Tampa Tribune,* April 12. http://web.-lexis-nexis.com/universe (accessed February 3, 2003).

———. 2001b. Aquifer Storage Bill Sinks after Public Outcry. *Tampa Tribune,* May 1. http://web.lexis-nexis.com/universe (accessed February 3, 2003).

Sarasota Herald-Tribune. 2001. Editorial. A deep-six treatment; flood of democracy stalled risky wells. May 7. http://web.lexis-nexis.com/universe (accessed February 2003).

Save Rodman Reservoir, Inc. (SRR). 2000. Governor Bush's Rodman Intentions. http://www.rodmanreservoir.com/Time_Line/bush.htm (accessed January 18, 2003).

Scarborough, Jerry. 2003. Speech of Jerry Scarborough, executive director, SRWMD, at American Farm Bureau Federation National Environmental Issues Round Table Suwannee River Partnership Presentation, May 29.

Schneider, Mark, John T. Scholz, Mark Lubell, Denisa Mindruta, and Matthew Edwardsen. 2003. Building Consensual Institutions: Networks and the National Estuary Program. *American Journal of Political Science* 47: 143–158.

Scholz, John T., and Mark Lubell. 1998. Trust and Taxpaying: Testing the Heuristic Approach to Collective Action. *American Journal of Political Science* 42: 398–417.

Schraeder-Freschette, Kristin S. 1993. *Burying Uncertainty: Risk and the Case Against Geological*

Disposal of Nuclear Waste. Berkeley, CA: University of California Press.

Seaton, Ned, and Jamal Thalji. 1996. Pinellas Going on Offense in Water Wars. *St. Petersburg Times,* January 17.

Seibold,Vince. 2003. Speech ofVince Seibold, FDEP, at Suwannee River Partnership Steering Committee Meeting, April 16.

Shaw, Margaret L. 1988. Mediator Qualifications: Report of a Symposium on Critical Issues in Dispute Resolution. *Seton Hall Legislative Journal* 12: 125–136.

Sloan, Mellini. 2003. Restoration of the Ocklawaha River and the Fate of the Rodman Reservoir. In *Adaptive Governance and Florida's Water Conflicts: The Case Studies,* edited by Aysin Dedekorkut, John Scholz, and Bruce Stiftel. Tallahassee FL: Florida State University, DeVoe L. Moore and Family Center for the Study of Economic Policy and Government, June, 161-177.

Smith, Darrell. 2002. The Suwannee River Partnership. *Poop Scoop*. University of Florida, Cooperative Extension Service, Institute of Food and Agricultural Sciences. 2002-01. http://nfrec-sv.ifas.ufl.edu/2002-01_poopscoop.htm (accessed July 1, 2003).

———. 2003a. Personal communication between Darrell Smith, coordinator, Suwannee River Partnership, and Aysin Dedekorkut. January 29.

———. 2003b. The 2002 Success Story. Suwannee River Partnership. http://mysuwanneeriver.com/resources/srp+2002+success.pdf (accessed February 16, 2004).

Smith, Rod, Evelyn J. Lynn, and Anna P. Cowin. 2003. Senate Bill 2042, George Kirkpatrick State Reserve.

Solochek, Jeffrey S. 2001. Geology Rules Out Aquifer Storage. *St. Petersburg Times*, April 18. http://web.lexis-nexis.com/universe. (accessed February 12, 2003).

Solow, Robert M. 2000. Sustainability: An Economist's Perspective. In *Economics of the Environment: Selected Readings*, 4th ed., edited by Robert M. Solow. New York: Norton.

South Florida Water Management District (SFWMD). 1997. Surface Water Improvement and Management (SWIM) Plan—Update for Lake Okeechobee. West Palm Beach, FL.

Southwest Florida Water Management District (SWFWMD). 2001. Partnership Agreement. *Issue Papers 2001*. August. http://www.swfwmd.state.fl.us/about/isspapers/partnershipagreement.html (accessed April 2, 2004).

St. Johns River Water Management District (SJRWMD). 2000. District Water Supply Plan. http://www.sjrwmd.org (accessed April 18, 2003).

———. 2003. East-Central Florida Water Supply Planning Initiative. Phase II. Annual Report of Activities and Accomplishments. http://www.sjrwmd.org. (accessed April 18, 2003).

———. No date. East Central Florida Water Use from the Floridan Aquifer. http://sjrwmd.com/programs/acq_restoration/res_devel/ecfla/wateruse_2020.jpg. (accessed April 18, 2003).

Stallworth, Holly. 2000. *Conservation Pricing of Waste and Wastewater.* Washington, DC: U.S. EPA.

Stavins, Robert N. 2003. Market-Based Environmental Policies: What Can We Learn from U.S. Experience (and Related Research)? Resources for the Future Discussion Paper 03-43. Washington, DC: Resources for the Future.

Stephens, Susan L. 2005. Florida's Waters: Significant Cases in the Works. Section Reporter, The Environmental and Land Use Law Section of the Florida Bar Association. March. http://www.eluls.org/2005/Reporter_Mar_2005/mar05_stephens.html (accessed May 23, 2005).

Stiftel, Bruce. 2000. Can Governments Bargain Effectively? Lessons from a Waste Transfer Station Location. Paper prepared for World Planning Schools Congress, Shanghai, China.

Stone, Deborah.1997. *Policy Paradox: The Art of Political Decision-Making.* New York: Norton.

Struhs, David B. 2001. Aquifer Storage Standards are the Focus. Letter to the editor. *Palm Beach Post,* April 14. http://web.lexis-nexis.com/universe (accessed February 12, 2003).

Struhs, David B., and Robert G. Brooks. 2001. Science behind ASR Technology is Sound. *In the News...* 3(16) April 20. http://www.myflorida.com/myflorida/governorsoffice/pdfs/newsletter-04-20-01.pdf (accessed February 07, 2004).

Sundqvist, Göran. 2002. *The Bedrock of Opinion.* Dordrecht: Kluwer Academic Publishers.

Susskind, Lawrence. 1981. Environmental Mediation and the Accountability Problem. *Vermont Law Review* 6(Spring): 1–47

Susskind, Lawrence, and J. Cruikshank. 1987. *Breaking the Impasse: Consensual Approaches to Resolving Public Disputes*. New York: Basic Books.

Susskind, Lawrence, S. McKearnan, and J. Thomas-Larmer (eds.). 1999. *The Consensus Building Handbook: A Comprehensive Guide to Reaching Agreement*. Thousand Oaks, CA: Sage Publications.

Sutherland, Donald. 1999. Florida's Sewage Could Undermine Billion-dollar Restoration Campaign. *Enviro-Net* 01. http://www.enviro-net.com (accessed February 14, 2004).

Suwannee River Partnership. 2002. *Reasonable Assurance Documentation for the Suwannee River, Santa Fe River and Suwannee River Estuary*. October 2. http://www.dep.state.fl.us/water/tmdl/docs/2002%20Update/suwrafinal.pdf (accessed February 16, 2004).

Suwannee River Water Management District (SRWMD). 2003a. Middle Suwannee Basin Work Plan. Draft. February 2.

———. 2003b. Protecting Florida's Groundwater through Best Management Practices. Video. http://mysuwanneeriver.com/features/suwannee+river+partnership/default1.htm (accessed February 16, 2004).

———. 2003c. *Suwannee River Partnership*. http://mysuwanneeriver.com/features/suwannee+river+partnership/default.htm (accessed February 28, 2004).

———. No date a. Programs and Research. http://mysuwanneeriver.com/features/suwannee+river+partnership/programs+and+research.htm (accessed February 16, 2004).

———. No date b. Suwannee River Partnership Cost Share Programs. http://mysuwanneeriver.com/resources/cost+share+programs.pdf (accessed February 16, 2004).

Swicegood, Ray. 2001. Improving and Protecting Water Quality in the Middle Suwannee River Area Watershed. Abstract from Natural Resources Forum—Watershed Science, Policy, Planning and Management: Can We Make It Work in Florida? June 19–21, Tampa, FL. http://conference.ifas.ufl.edu/nrf/abstract.pdf (accessed February 16, 2004).

Swichtenberg, Bill. 2001. Tampa Desalination Plant Faces Suit. *Water Engineering and Management* 148(11): 8.

Tampa Bay Soundings. 2003. Desal Comes On Line. Spring (2). http://www.baysoundings.com/spring03/briefs.html#desal (accessed April 2, 2004).

Tampa Bay Water. 2001. Administrative Law Judge Recommends DEP Issue Proposed Permit for Desalination Plant on Tampa Bay. October 17. http://www.tampabaywater.org/WEB/Htm/News/news-item27.htm (accessed April 2, 2004).

———. No date. About Us. http://www.tampabaywater.org/WEB/Htm/About-Us (accessed April 2, 2004).

Thomas, Elizabeth. 2003. Personal communication from Elizabeth Thomas, Senior Project Manager. Division of Water Supply Management, SJRWMD to Aysin Dedekorkut. March 11.

Thomas, Hugh. 2003. Speech of Hugh Thomas, coordinator, Suwannee River Partnership Santa Fe Basin Expansion, at Suwannee River Partnership Steering Committee Meeting, January 29.

U.S. Army Corps of Engineers (USACE). 1968. *Water Resources for Central and Southern Florida*. Jacksonville, FL: Jacksonville District, USACE.

———. 1989. Post-Authorization Change Notification Report for the Reallocation of Storage from Hydropower to Water Supply at Lake Lanier, Georgia. Mobile, AL: Mobile District, USACE.

———. 1994. *Central and Southern Florida Project, Comprehensive Review Study, Reconnaissance Report*. (November). Jacksonville, FL: Jacksonville District, USACE.

———. 1996. *Florida's Everglades Program, Everglades Construction Project, Final Programmatic Environmental Impact Statement*. (September). Jacksonville, FL: Jacksonville District, USACE.

———. 1998. *Central and Southern Florida Project Comprehensive Review Study, Draft Integrated Feasibility Report and Programmatic Environmental Impact Statement*. Jacksonville, FL: Jacksonville District, USACE.

U.S. Department of Agriculture, Forest Service. 2002a. *Environmental Impact Statement*. Tallahassee: USDA/FS.

———. 2002b. Special Use Permit for Occupation of National Forest System Lands by the Office of Greenways and Trails, Florida DEP. Tallahassee: USDA/FS.

U.S. Department of Interior, U.S. Department of Agriculture, U.S. Department of the Army,

U.S. Department of Commerce, U.S. Department of Justice, and U.S. Environmental Protection Agency. 1993. *Interagency Agreement on South Florida Ecosystem Restoration.* September 23. Washington, DC: U.S. Department of Interior. http://www.sfrestore.org/documents/interagency_agreement_1993.pdf (accessed May 19, 2005).

U.S. Environmental Protection Agency (EPA). 1998. Letter from John Hankinson, Regional Administrator, to Ernie E. Frey, Director of District Management, FDEP North District, Jacksonville. 4WM-SWPFB/EAD. March 26.

———. 2000. Revision to the Federal Underground Injection Control (UIC) Requirements for Class I Municipal Wells in Florida. 65 *Federal Register* 131: 2000.

———. 2003. Drinking Water State Revolving Loan Program, Financing America's Drinking Water from the Source to the Tap. Report to Congress. EPA 918-R-03-009, May.

U.S. Executive Office of the President. 2000. *Economic Report of the President 2000.* Washington, DC: U.S. Government Printing Office.

U.S. Fish and Wildlife Service. 1998. *Draft Fish and Wildlife Coordination Act Report, Central and Southern Florida Comprehensive Review Study.* Vero Beach, FL: South Florida Restoration Office.

U.S. Geological Survey (USGS). 2003. National Water Conditions. February 5. http://water.usgs.gov/nwc/explain_data.html#wateryear (accessed February 16, 2004).

U.S. Government Accounting Office (GAO). 1997. *Regulatory Reinvention: EPA's Common Sense Initiative Needs an Improved Operating Framework and Progress Measures.* Washington, DC: GAO, 97-164.

Vasishth, Ashwani. 1996. Think Globally, Act Locally?: An Ecosystem Approach to Scale in Environmental Planning. Paper presented at the Joint International Congress of the Association of Collegiate Schools of Planning and the Association of the European Schools of Planning, July 25–28, Toronto, Ontario.

Vecchioli, John. 1999. *Aquifer Storage and Recovery: Is it South Florida's Solution for Regional Water Storage?* Presentation at South Florida Restoration Science Forum, May 17–19, Boca Raton, FL. http://sofia.usgs.gov/sfrsf/presentations/vecchioli.html (accessed May 20, 2003).

———. 2003. Personal communication from John Vecchioli, retired USGS district chief, to Eberhard Roeder. December 10.

Vergara, Barbara. 1998. Water Supply Assessment 1998. Executive Summary. http://sjrwmd.com/programs/plan_monitor/sup_planning/98ns_exs.html (accessed April 11, 2003).

———. 2000. District Water Supply Plan. Executive Summary. http://www.sjrwmd.org (accessed April 18, 2003).

———. 2003. Personal communication to Ramiro Berardo from Barbara Vergara, Director, Division of Water Supply Management, SJRWMD. March 13.

Wanless, Harold. 2003. Presentation given by Harold Wanless, University of Miami geology professor, at NGWA National Exposition and Convention, December 10, Orlando, FL.

Water Governance Working Group. 2003. Apalachicola-Chattahoochee-Flint Tri-State Water Conflict. In *Adaptive Governance and Florida's Water Conflicts: The Case Studies*, edited by Aysin Dedekorkut, John Scholz and Bruce Stiftel. Tallahassee: Florida State University, DeVoe L. Moore and Family Center for the Study of Economic Policy and Government, 179-186.

Webster, Kirk. 2003. Personal communication to Aysin Dedekorkut from Kirk Webster, Deputy Executive Director, SRWMD, January 29.

Weible, Chris, and Paul Sabatier. 2005. Comparing Policy Networks: Marine Protected Areas in California. *Policy Studies Journal* 33(2): 181-201.

Weimer, David L., and Aidan R. Vining. 1999. *Policy Analysis: Concepts and Practice.* 3rd ed. Englewood Cliffs, NJ: Prentice-Hall.

Weinberg, Alvin. 1972. Science and Trans-science. *Minerva* 10: 209–222.

Weiss, Carol. 1977. Research for Policy's Sake: The Enlightenment Function of Social Research. *Policy Analysis* 3(Fall): 531–545.

Wengert, Norman. 1971. Public Participation in Water Planning: A Critique of Theory, Doctrine and Practice. *Water Resources Bulletin* 7: 26–32.

Wilhere, G.F. 2002. Adaptive Management in Habitat Conservation Plans. *Conservation Biology* 16(1): 20–29

Wondolleck, Julia Marie, and Steven L. Yaffee. 2000. *Making Collaboration Work: Lessons from Innovation in Natural Resource Management*. Washington, DC: Island Press.

Woods, Chuck. 2001. UF Research Targets Rising Nitrates in Suwannee River Basin. *UF News* February 26. http://www.napa.ufl.edu/2001news/suwannee.htm (accessed February 16, 2004).

The World Commission on Environment and Development. 1987. *Our Common Future*. Oxford, England: Oxford University Press.

York, David. 2003. Personal communication from David York, FDEP Water Reuse program manager, to Eberhard Roeder. March 26.

Young, Robert A. 2005. *Determining the Economic Value of Water: Concepts and Methods*. Washington, DC: Resources for the Future.

Zafonte, Matthew, and Paul Sabatier. 1998. Shared Beliefs and Imposed Interdependencies as Determinants of Ally Networks in Overlapping Subsystems. *Journal of Theoretical Politics* 10: 473–505.

Zartman, William. 1991. Conflict and Resolution. *Annals of the American Academy of Conflict Resolution* 518(November): 11–20.

List of Abbreviations

ACF	Advocacy Coalition Framework
ACF	Apalachicola-Chattahoochee-Flint
ACT	Alabama-Coosa-Tallapoosa
ACSC	Areas of Critical State Concern
ADECA	Alabama Department of Economic and Community Affairs
ADR	Alternate Dispute Resolution
ALJ	Administrative Law Judge
APA	Administrative Procedures Act
ASR	Aquifer Storage and Recovery
AWWA	American Water Works Association
BATNA	best alternative to a negotiated agreement
BMPs	best management practices
CAFO	Concentrated Animal Feeding Operations
CARES	County Alliance for Responsible Environmental Stewardship
CEQ	Council on Environmental Quality
CERP	Comprehensive Everglades Restoration Plan
CFBC	Cross-Florida Barge Canal
CLAC	Canal Lands Advisory Committee
COLA	The Coalition of Lake Associations
CROGEE	Committee on the Restoration of the Greater Everglades Ecosystem
CSI	Common Sense Initiative
CUP	Consumptive Use Permit
CWA	Clean Water Act
DACS	Department of Agriculture and Consumer Services

DCA	Department of Community Affairs
DEP	Department of Environmental Protection
DOAH	Florida Division of Administrative Hearings
DOH	Florida Department of Health
DOI	U.S. Department of the Interior
DNR	Florida Department of Natural Resources
DRI	Developments of Regional Impact
DWSP	District Water Supply Plan
EIS	Environmental Impact Statements
EPA	U.S. Environmental Protection Agency
EQIP	Environmental Quality Incentive Program
ERP	Environmental Resource Permitting
FCRC	Florida Conflict Resolution Consortium
FDACS	Florida Department of Agriculture and Consumer Services
FDE	Florida Defenders of the Environment
FDEP	Florida Department of Environmental Protection
FDER	Florida Department of Environmental Regulation
FDOH	Florida Department of Health
FFWCC	Florida Fish and Wildlife Conservation Commission
FGFWFC	Florida Game and Freshwater Fish Commission
Fla. Admin. Code	Florida Administrative Code
Fla. Stat.	Florida Statutes
FNRLI	Florida Natural Resources Leadership Institute
FWS	Fish and Wildlife Service
HOPE	Help Our Polluted Environment
IFAS	University of Florida's Institute of Food and Agricultural Services
IWR	Impaired Waters Rule
LOPA	Lake Okeechobee Protection Act
LTCs	Long Term Contracts
MFL	minimum flows and levels
mgd	million gallons per day
mg/L	milligrams per liter
MIL	The Mobile Irrigation Laboratory
MOA	Memorandum of Agreement
MOU	Memorandum of Understanding
MSRA	Middle Suwannee River Area
n.d.	No date
NEPA	National Environmental Policy Act

NMFS	National Marine Fisheries Service
NOAA	National Oceanic and Atmospheric Administration
NPDES	National Pollutant Discharge Elimination System
NPS	National Park Service
NRCS	Natural Resources Conservation Service
OECD	Organization for Economic Cooperation and Development
P&G	Proctor and Gamble
PL	Public Law
PRO	Proposed Recommended Order
PSC	Public Service Commission
PWRCA	Priority Water Resource Caution Areas
RWSA	Regional Water Supply Authority
SFWMD	South Florida Water Management District
SJRWMD	St. Johns River Water Management District
SLAPP	Strategic Lawsuit Against Public Participation
SOBAC	Save Our Bays and Canals
SRR	Save Rodman Reservoir, Inc.
SRWMD	Suwannee River Water Management District
SUP	Special Use Permit
SWFWMD	Southwest Florida Water Management District
SWIM	Surface Water Improvement and Management
TMDL	Total Maximum Daily Loads
TNC	The Nature Conservancy
UF	University of Florida
USACE	U.S. Army Corps of Engineers
USDA	United States Department of Agriculture
USDA/FS	U.S. Department of Agriculture's Forest Service
USGS	U.S. Geological Survey
WARN	Water Assessment Regional Network
WCRWSA	West Coast Regional Water Supply Authority
WMD	Water Management District
WQBEL	Water Quality Based Effluent Limitations
WSFWP	Water Supply Facilities Work Plan

Index

Page numbers in italics refer to figures.